Printed Antennas for Future Generation Wireless Communication and Healthcare

This book focuses on the design and development of printed antennas along with modeling aspects for multifaceted applications. It further investigates imperfections in the manufacturing processes and assembly operation during the testing/characterization of printed antennas.

This text:
- Discusses in a comprehensive manner the design and development aspects of printed antennas.
- Provides fractal engineering aspects for miniaturization and wideband characteristics of low-profile antennas with high performance.
- Covers high-gain printed antennas for Terahertz application.
- Showcases electrical modeling of smart antennas.
- Pedagogical features such as review questions based on practical experience are included at the end of each chapter.

The book comprehensively discusses fractal engineering in printed antennas for miniaturization and enhancement of performance factors. It further covers the modeling of electrically small antennas, circuit modeling, modeling of factual-based ultra-wide band antennas, and modeling of reconfigurable micro-electromechanical system-based patch antennas. The book highlights performance metrics of multiple-input–multiple-output antennas. It will serve as an ideal reference text for senior undergraduate and graduate students and academic researchers in fields including electrical engineering, electronics, communications engineering, and computer engineering.

Printed Antennas for Future Generation Wireless Communication and Healthcare

Balaka Biswas and Ayan Karmakar

CRC Press
Taylor & Francis Group
Boca Raton London

CRC Press is an imprint of the
Taylor & Francis Group, an **informa** business

First edition published 2023
by CRC Press
6000 Broken Sound Parkway NW, Suite 300, Boca Raton, FL 33487–2742

and by CRC Press
4 Park Square, Milton Park, Abingdon, Oxon, OX14 4RN

CRC Press is an imprint of Taylor & Francis Group, LLC

© 2023 Balaka Biswas and Ayan Karmakar

Reasonable efforts have been made to publish reliable data and information, but the author and publisher cannot assume responsibility for the validity of all materials or the consequences of their use. The authors and publishers have attempted to trace the copyright holders of all material reproduced in this publication and apologize to copyright holders if permission to publish in this form has not been obtained. If any copyright material has not been acknowledged please write and let us know so we may rectify in any future reprint.

Except as permitted under U.S. Copyright Law, no part of this book may be reprinted, reproduced, transmitted, or utilized in any form by any electronic, mechanical, or other means, now known or hereafter invented, including photocopying, microfilming, and recording, or in any information storage or retrieval system, without written permission from the publishers.

For permission to photocopy or use material electronically from this work, access www.copyright.com or contact the Copyright Clearance Center, Inc. (CCC), 222 Rosewood Drive, Danvers, MA 01923, 978–750–8400. For works that are not available on CCC please contact mpkbookspermissions@tandf.co.uk

Trademark notice: Product or corporate names may be trademarks or registered trademarks and are used only for identification and explanation without intent to infringe.

Library of Congress Cataloging-in-Publication Data
Names: Biswas, Balaka, author. | Karmakar, Ayan, author.
Title: Printed antennas for future generation wireless communication
   and healthcare / Balaka Biswas and Ayan Karmakar.
Description: First edition. | Boca Raton : CRC Press, [2023] | Includes
   bibliographical references and index.
Identifiers: LCCN 2022055118 (print) | LCCN 2022055119 (ebook)
Subjects: LCSH: Microstrip antennas. | Printed circuits. | Wireless communication
   systems—Equipment and supplies. | Microwave imaging in medicine.
Classification: LCC TK7871.67.M5 B57 2023 (print) | LCC TK7871.67.M5
   (ebook) | DDC 621.382/4—dc23/eng/20230111
LC record available at https://lccn.loc.gov/2022055118
LC ebook record available at https://lccn.loc.gov/2022055119

ISBN: 978-1-032-39301-8 (hbk)
ISBN: 978-1-032-48604-8 (pbk)
ISBN: 978-1-003-38985-9 (ebk)

DOI: 10.1201/9781003389859

Typeset in Sabon
by Apex CoVantage, LLC

Contents

Acknowledgments	xi
Foreword	xiii
Preface	xv
About the Authors	xvii
List of Abbreviations and Acronyms	xix

1 Printed Antennas for Modern-Day Communication   **1**

    1.1   Introduction 1

    1.2   Types of Printed Antennas 1

        1.2.1   Microstrip Antennas 1

        1.2.2   Slot Antennas 2

        1.2.3   Dipole Antennas 3

        1.2.4   Monopole Antennas 3

        1.2.5   Inverted F-Antennas 5

        1.2.6   Log Periodic Antennas 5

        1.2.7   Quasi-Yagi Antennas 6

        1.2.8   Fractal Antennas 7

        1.2.9   Bow Tie Antennas 7

        1.2.10 Spiral Antennas 8

        1.2.11 Microstrip Leaky Wave Antennas 8

    1.3   Summary 10

    References 10

    Problems 12

2 Fractals in Printed Antennas   **15**

    2.1   Introduction 15

    2.2   Cantor Set 16

    2.3   Koch Curve 16

    2.4   Sierpinski Carpet 17

vi Contents

2.5 *Sierpinski Gasket 18*
2.6 *Minkowski Fractal 19*
2.7 *Pythagoras Tree Fractal 19*
2.8 *Hilbert Curve Fractal 19*
2.9 *Background Study of Fractals in Antenna Engineering 19*
    2.9.1 *UWB Fractal Antennas With Elliptical/Circular Patches 20*
    2.9.2 *UWB Fractal Antennas With Square/Rectangular Patch 21*
    2.9.3 *UWB Fractal Antennas of Different Shapes 22*
    2.9.4 *UWB Fractal Antennas With Ground Plane 23*
2.10 *Recent Works on Fractals in Antenna Engineering 24*
2.11 *Summary 29*
References 31
Problems 33

3 Journey of UWB Antennas Towards Miniaturization   35

3.1 *Introduction 35*
3.2 *Initial Developmental Phase 36*
3.3 *Improvement/Developmental Phase of UWB Antennas 38*
    3.3.1 *UWB Planar Metal Plate Monopole Antennas 38*
    3.3.2 *UWB Printed Monopole Antennas 40*
        3.3.2.1 *Microstrip Fed Printed Monopole Antennas 41*
        3.3.2.2 *Coplanar Waveguide Fed Printed Monopole Antennas 42*
    3.3.3 *UWB Slot/Aperture Monopole Antennas 43*
        3.3.3.1 *Microstrip Fed Slot/Aperture Monopole Antennas 43*
        3.3.3.2 *CPW Fed Slot/Aperture Monopole Antennas 46*
3.4 *Band Notched Characteristics of UWB Antennas 48*
    3.4.1 *Single-Band Notched UWB Antennas 48*
        3.4.1.1 *Slots on the Radiating Patch 48*
        3.4.1.2 *Parasitic Strips in UWB Slot Antennas 52*
        3.4.1.3 *Slots on the Ground Plane 52*
        3.4.1.4 *Slots on Feed Line 54*
    3.4.2 *Dual-Band Notched UWB Antennas 54*
    3.4.3 *Multiple-Band Notched UWB Antennas 55*

Contents   vii

3.5   *A Brief Review of Tapered Slot Antennas 59*
        3.5.1   *Compact Design of Vivaldi Antennas 62*
        3.5.2   *Improved Antipodal Vivaldi Antennas 64*
3.6   *Summary 65*
References 66
Problems 71

4 Modeling of Printed Antennas   73

4.1   *Introduction 73*
4.2   *Modeling of Electrically Small Antennas 73*
4.3   *Circuit Modeling 74*
        4.3.1   *Antenna I 74*
        4.3.2   *Antenna II 77*
        4.3.3   *Antenna III 79*
4.4   *Modeling of Fractal-Based UWB Antennas 83*
        4.4.1   *Antenna I 84*
        4.4.2   *Antenna II 86*
        4.4.3   *Antenna III 88*
4.5   *Modeling of Reconfigurable MEMS-Based*
        Patch Antennas 90
4.6   *Circuit Modeling 92*
4.7   *Summary 95*
References 95
Problems 98

5 Printed Antennas for Biomedical Applications   99

5.1   *Introduction 99*
5.2   *Antenna Development Inside the Capsule 100*
        5.2.1   *Antenna Design 100*
        5.2.2   *Parametric Studies 102*
        5.2.3   *Measurement of the Antenna 109*
5.3   *Antenna Development Outside the Capsule 116*
        5.3.1   *Antenna Design 116*
        5.3.2   *Electrical Equivalent Circuit 122*
        5.3.3   *Measurement of Antennas 125*
5.4   *Link Budget Analysis 126*
5.5   *Summary 129*
References 129
Problems 131

viii   Contents

6   **High-Gain Printed Antennas for Sub-Millimeter Wave Applications**   **133**

6.1   *Introduction 133*

6.2   *Design and Development of High-Gain Printed Antennas 133*

    6.2.1   *Design of Parasitic Element Loaded Microstrip Antenna Arrays With Enriched Gain 134*

    6.2.2   *Antenna Fabrication 137*

6.3   *Summary 139*

References 139

Problems 140

7   **Systematic Investigation of Various Common Imperfections in Printed Antenna Technology and Empirical Modeling**   **141**

7.1   *Introduction 141*

7.2   *Fabrication-Related Imperfections 141*

    7.2.1   *Surface Roughness or Hillocks 141*

        7.2.1.1   *Hammerstad-Bekkadal Model 142*

        7.2.1.2   *Huray Model 143*

    7.2.2   *Voids in Via-Hole Filling 145*

    7.2.3   *Voids in Traditional Conducting Plates of Radiator or Feed Networks 145*

    7.2.4   *Unexpected Polymer Residues in the Gap Between Two Metal Traces 146*

    7.2.5   *Cavity Opening Problem in Bulk Micro-Machined Antennas 146*

    7.2.6   *Sagging Problems in Surface Micromachining 148*

    7.2.7   *Improper Wafer Bonding or Misalignment in 3D Vertical Integration 148*

7.3   *IC Assembly or Packaging-Related Issues 150*

    7.3.1   *Effect of Epoxy Spreading 150*

    7.3.2   *Improper Wire Bonding 150*

7.4   *Summary 152*

References 153

Problems 154

8   **Multiple Input and Multiple Output Antennas**   **155**

8.1   *Introduction 155*

8.2   *Performance Metrics of MIMO Antennas 158*

    8.2.1   *Envelope Correlation Coefficient 159*

Contents ix

8.2.2   *Diversity Gain 159*
8.2.3   *Total Active Refection Coefficient 159*
8.2.4   *Channel Capacity Loss 160*
8.3   *Challenges in MIMO Antenna Design 160*
8.4   *Different Methods to Reduce Mutual Coupling 160*
  8.4.1   *Defected Ground Structure 161*
  8.4.2   *Electromagnetic Band Gap Structures 163*
  8.4.3   *Neutralization Line 166*
  8.4.4   *Metamaterial Structures 170*
  8.4.5   *Parasitic Elements 177*
8.5   *Types of MIMO Antennas 177*
  8.5.1   *Single-Band MIMO Antennas 178*
  8.5.2   *Dual/Multiband MIMO Antennas 178*
  8.5.3   *Wide-Band MIMO Antennas 179*
  8.5.4   *Ultra-Wideband MIMO Antennas 183*
8.6   *MIMO Antennas in Biomedical Usage 188*
8.7   *Reconfigurable MIMO Antennas 188*
8.8   *Summary 192*
References 192
Problems 196

9   Antennas for Microwave Imaging   199

9.1   *Introduction 199*
  9.1.2   *Why Microwave? 199*
  9.1.3   *Challenges to Designing Microwave Imaging Systems 200*
  9.1.4   *Application of Microwave Imaging 202*
9.2   *Types of Microwave Imaging 203*
9.3   *Antennas for Microwave Imaging 204*
  9.3.1   *Medical Imaging Applications 204*
    9.3.1.1   *Monopole Antennas 205*
    9.3.1.2   *Vivaldi Antennas 206*
    9.3.1.3   *Bow-Tie Antennas 209*
    9.3.1.4   *Fractal Antennas 214*
    9.3.1.5   *Horn Antenna 216*
  9.3.2   *Concealed Weapon Detection Applications 220*
  9.3.3   *Structural Health Monitoring 221*
9.4   *Summary 221*
References 221
Problems 224

10 Rectennas: A New Frontier for Future Wireless Communication   225

10.1   Introduction 225
10.2   Performance Metrics of RF Harvesters 226
    10.2.1   Range of Operation 227
    10.2.2   Power Conversion Efficiency of RF to DC 228
    10.2.3   Q-Factor of the Resonator 229
    10.2.4   Sensitivity 230
    10.2.5   Output Power 230
10.3   Design Protocol of RF Energy Harvesting Circuits 230
10.4   Building Blocks of Rectennas 231
    10.4.1   Front-End Antennas 231
    10.4.2   Impedance Matching Networks 235
    10.4.3   Rectifying Circuits 235
10.5   Practical Applications of RF Energy Harvesters 237
    10.5.1   Healthcare Devices/Biomedical Engineering 237
    10.5.2   Future Wireless Communication Networks 237
    10.5.3   Explosive Detection Missions 238
    10.5.4   Unique Solutions for Specific Military Applications 238
10.6   Summary 238
References 239
Problems 241

Future Scope   243
Appendices   245
Appendix-1   RF And Microwave Frequency Spectrum Along With Practical Applications 245
Appendix-2   Comparison of Different Planar Antenna 246
Appendix-3   Comparison of Various Computational Electromagnetic Solvers 246
Appendix-4   Popular Types of Planar Antennas for Antenna-in-Package (AiP) Configuration 247

Index   249

Acknowledgments

It is our privilege and pleasure to express our profound senses of deep respect, sincere gratitude, and thankfulness to The ALMIGHTY without whose blessings and continuous motivation this mammoth task could not have been accomplished. Our parents and their inspiring words in every corner of our lives have made this thing materialize. Numerous hurdles came during this journey, but ultimately positive thinking and strong willpower win the race. We are thankful to all the well-wishers for compelling us to think positively and overcoming various challenges encountered during this work. The comments and suggestions during various stages of this manuscript improved the quality, and we wish to acknowledge all the reviewers for their effort.

This book would not have been written without the encouragement of and interaction with many people. We thank our friends in the industry and academia for their support on projects and applications. Much of the information contained in this book is the result of our research and investigation on printed antennas carried out at SCL, Chandigarh, Jadavpur University, Kolkata, NIT-Durgapur and CSIO Lab in Chandigarh. Therefore, we take this opportunity to thank all the technologists of CSIR-CSIO, Chandigarh, SCL-Chandigarh, and academicians from JU and NIT-DGP, who gave unconditional support at every stage whenever desired. We especially thank director-SCL, Mr. Amitava Das, senior principal scientist of CSIO, and Prof. D.R. Poddar of JU and Prof. Rowdra Ghatak of NIT-DGP who gave occasional guidance in this research work.

Our grateful thanks are due to Dr. Gauravjeet Singh Reen and his team at CRC Press, Taylor and Francis Group, for their unstinting support and cooperation in bringing out this edition, which will be beneficial for students and academia.

For the time that has been stolen from our families, we acknowledge the support of all family members, especially from our son, Avranil, and we will continue to seek forgiveness from them.

Finally, we will appreciate and gratefully acknowledge any constructive comments, suggestions, and criticisms from any reader of our book.

Balaka Biswas
Ayan Karmakar
Chandigarh, India

Foreword

There has been enormous growth in printed antenna technology for future-generation wireless communication and healthcare services, covering radio-frequency (RF) and microwave gametes. Efforts are already underway for extending the same technology to sub-THz (Terahertz) waves. THz technology is also knocking at the door of antenna engineers with tremendous potential in miniaturization and promising features for strategic applications.

This book is a practical antenna design guide, covering various real-world applications for those who are plunging into this field. It blends proportionately insight theory and practical design issues of antenna engineering in a beautiful manner. It stems from the extensive experiences of the authors in the analysis, design, and development of various miniaturized printed antennas for future-generation wireless communication and biomedical applications. The profound knowledge of both authors on the subject makes this book an authoritative contribution in the field.

Usually, design engineers do not have adequate opportunity to teach, and academicians rarely have the experience of developing devices and systems to the ultimate point of seeing them tested in the field and accepted in service. The authors have combined these two distinct experiences in a very unusual manner.

I wish to congratulate the authors for their efforts in compiling various topics into one book, and I am sure that, with such a background and variety of experience, this book will be of great value both as a textbook for graduate courses and a reference book for practical antenna engineers and technicians.

Prof. (Dr.) Deepak Ranjan Poddar
Ex-Emeritus Professor, Department of ETCE,
Jadavpur University, Kolkata, India
Fellow Member of IEEE
Life Fellow of the IE and IETE
Life Member of the Society of EMC Engineers

Preface

The purpose of this book is to provide a practical antenna design guide covering various real-world applications for those who are plunging into this area. There are already many excellent category textbooks on antenna theory, mathematical formulations, and design. But this book focuses more on practical antenna design and development rather than only theoretical analysis. It tries to find harmony between the analytical approach and practical implementations of antennas. In industry and academia, it is very hard to get a general understanding of the field prior to being engaged in a specialized field. Engineers who are engaged in the field of wireless communication, the aerospace or healthcare industries, 5G, and many other areas will find this book useful.

Senior antenna engineers and managers can use this book as training material for new hires. Undergraduate, postgraduate, and PhD scholars can use it to quickly gain expert-level proficiency. We believe that this concise book will be a welcome addition to the collection of books on this subject.

We have tried to organize the contents in a flow starting from an overview of various modern-day printed antennas, miniaturization techniques, and electrical equivalent circuit modeling of many small antennas and covering two rising fields of printed antennas and finally investigating various aspects of antenna fabrication.

Chapter 1 discusses various types of printed antennas with their practical applications. It covers a large family of antennas used in several fields for modern-day wireless communication.

Chapter 2 touches upon one of the most popular techniques in the printed antenna field for miniaturization and enhancement of performance factors, fractal engineering.

Chapter 3 gives a concise literature survey along with the work of the authors' group on ultra-wideband (UWB) antennas towards miniaturization. It sketches an outline for the journey of UWB antennas targeting multifunctional applications.

Chapter 4 explains the modeling aspect of various printed antennas. A simple, easy-to-understand approach has been adopted to build electrical equivalent circuits for several antennas.

Chapter 5 outlines the development of printed antennas for biomedical applications. The wireless aspect endoscopy application is especially targeted here.

Chapter 6 describes the design and development work on printed antennas for sub-millimeter wave applications with enriched gain characteristics. Moreover, this chapter deals with antennas on flexible substrate so that this kind of antenna can be implemented practically in a conformal shape.

Chapter 7 deals with the systematic investigation of various common imperfections involved in the manufacturing and assembly process of printed antennas. Empirical modeling of each problem is elaborated.

Chapter 8 elaborates on the need of multi-input–multi-output (MIMO) antennas in the wireless communication field, and then a thorough survey of recent works in the context of printed versions of this kind of antenna is summarized. The need for such antennas in modern-day biomedical engineering is also explored.

Chapter 9 outlines the state-of-the art work on printed antennas for various medical imaging purposes. UWB antennas play a crucial role in this task.

Finally, Chapter 10 introduces a new kind of printed antenna along with rectifying circuits called rectenna targeting wireless power transfer (WPT). WPT is a prime choice for future wireless sensor networks and biomedical engineering, and it has its own importance.

Last, the future scope highlights the opportunity for wide applications of printed antennas for various aspects of human life.

About the Authors

Dr. Balaka Biswas received her BE in electronics and communication engineering from the University of Burdwan, West Bengal, and thereafter finished her master's in the year 2008. She received 1st class 1st (gold medal) in MTech and subsequently joined the PhD program at Jadavpur University, Kolkata, with a DST-INSPIRE fellowship. Later, she joined CSIO-CSIR, Chandigarh, as a SRA under the Scientist Pool scheme and did her post-doc from IISER, Mohali. Currently, she is working at Chandigarh University, Mohali, as an associate professor. She has more than 25 publications in reputed referred journals and conferences. She serves as a reviewer in several top-tier journals, such as *Institute of Electrical and Electronics Engineers Transactions on Antenna Propagation*, *The Institution of Electronics and Telecommunication Engineers* journals, *PhotonIcs and Electromagnetics Research-C*, *PhotonIcs and Electromagnetics Research-M*, *Journal of Electromagnetic Waves and Applications*, PIER-C, Microwave and Optical technology letters, IEEE access and others. She is a life member of IETE, ISSE, and Indian Science Congress Association.

Ayan Karmakar received a BTech degree (electronics and communication) from WBUT, Kolkata, in 2005. Later he did his master's at NIT, Durgapur. He joined ISRO as a scientist and subsequently posted to the Semi-Conductor Laboratory (SCL), Chandigarh. His research interests include design and development of passive microwave integrated circuits and antennas using silicon-based MIC and RF-MEMS technology. His field of expertise is expanded in fabrication technology of various MEMS-based sensors and evolving test strategies for the same.

He has more than 40 publications in reputed journals and conferences at various national and international levels. In 2019, he authored one technical book, *Si-RF Technology*, from Springer, Singapore. He is a fellow member of IETE, India, and life member of the Indian Science Congress Association, ISSE, and Bangiya Bigyan Parishad.

Abbreviations and Acronyms

ADS	Advanced Design System
AiP	Antenna-in-Package
ASIC	Application Specific Integrated Circuit
AUT	Antenna Under Test
AVA	Antipodal Vivaldi Antenna
CLL	Capacitively Loaded Loop
CMOS	Complementary Metal Oxide Semiconductor
C/N$_0$	Carrier-to-Noise Ratio
COB	Chip on Board
CPW	Co-Planar Waveguide
CWSA	Constant Width Slot Antenna
DC	Direct Current
DCS	Distributed Control System
DETSA	Dual Elliptically Tapered Antipodal Slot Antenna
DGBE	Diethylene Glycol Butyl Ether
EBG	Electromagnetic Band Gap
EIRP	Effective Isotropic Radiated Power
EM	Electromagnetic
E-Plane	Elevation Plane
ESA	Electrically Small Antenna
FCC	Federal Communication Commission
FDTD	Finite Difference Time Domain
FEM	Finite Element Method
FIT	Finite Integrated Technique
FOM	Figure of Merit
FR-4	Flame Retardant 4
5G	Fifth-Generation Wireless/Mobile Communication
GHz	Gigahertz
GI	Gastrointestinal Tract
GPS	Global Positioning System
G/T	Gain-to-Noise Ratio
HFSS	High-Frequency Structure Simulator

H-Plane	Horizontal/Azimuth Plane
HRS	High Resistive Silicon
IC	Integrated Chip
IEEE	Institution of Electrical and Electronics Engineering
IFA	Integrated-F-Antenna
IoT	Internet of Things
ISM	Industrial, Scientific, and Medical
ITU-T	International Telecommunications Union
Ka-band	IEEE Radar Band Letter Designation for 26.5–40 GHz
KOH	Potassium Hydroxide
KSIR	Kirchhoff's Surface Integral Representation
Ku-band	IEEE Radar Band Letter Designation for 18–26.5 GHz
LCP	Liquid Crystal Polymer
LED	Light-Emitting Diode
LM	Link Margin
LPDA	Log Periodic Dipole Antenna
LTSA	Linearly Tapered Slot Antenna
LTE	Long-Term Evolution
MCM	Multichip Module
MEMS	Micro-Electro Mechanical System
MHz	Megahertz
MICS	Medical Implant Communication Services
MIM	Metal Insulator Metal
MIR	Micro Power Impulse Radar
MIMO	Multiple Input–Multiple Output
MMIC	Monolithic Microwave Integrated Circuit
mmW	Millimeter Wave
MSL	Micro Strip Line
NaCL	Sodium Chloride
OCA	On-Chip-Antenna
OFDM	Orthogonal Frequency Division Multiplexing
PCB	Printed Circuit Board
PCS	Process Control System
PEC	Perfect Electric Conductor
PEEK	Polyetherether Ketones
PGK	Penta-Gasket-Khoch
PTMA	Printed Tapered Monopole Antenna
Q-factor	Quality Factor
RADAR	Radio Detection and Ranging
RF	Radio Frequency
RMAA	Rectangular Microstrip Antenna Array
SAR	Specific Absorption Rate
SiO_2	Silicon Dioxide

SIR	Stepped Impedance Resonator
SoC	System on Chip
SOLT	Short Open Load Through
SRR	Split Ring Resonator
SWB	Super Wide Band
THz	Terahertz
TM	Transverse Magnetic Field
TMAH	Tetra Methyl Amino Hydroxide
TOSM	Through Open Short Match
TRL	Through Reflection Load
T/R	Transmitter/Receiver
TSA	Tapered Slot Antenna
3D	Three Dimensional
UMTS	Universal Mobile Telecommunications System
UWB	Ultra-Wideband
VLSI	Very Large-Scale Integration
VNA	Vector Network Analyzer
VSWR	Voltage Standing Wave Ratio
WBAN	Wireless Body Area Network
WCE	Wireless Capsule Endoscopy
WiFi	Wireless Fidelity
WiMAX	Worldwide Interoperability for Microwave Access
WLAN	Wireless Local Area Network

Chapter 1

Printed Antennas for Modern-Day Communication

1.1 INTRODUCTION

Antennas are the building blocks of any communication system. An antenna is a transducer which converts an electrical signal into an electromagnetic signal in free space (transmitting antenna) and vice versa (receiving antenna). Antenna engineers are continuously trying to improve according to our modern communication needs that demand compact, light weight, portable devices for multiband utilization with excellent features and a low profile that are easy to manufacture with cost-effective manufacturing process. Antenna can be widely used in military, commercial, medical, and space applications; mobile communications; global positioning systems (GPSs); and so on. Traditionally printed antennas can serve all these purposes due to their excellent features along with good performance characteristics, durability, and compactness that make them suitable to integrate easily with microwave/millimeter wave circuit components in the same PCB using sophisticated tools.

1.2 TYPES OF PRINTED ANTENNAS

A printed antenna is one type of planar antenna that allows antennas to be used in applications where size and shape are crucial. This flexibility of application has led to the development of a wide variety of planar antenna classes. Here attempts are made to classify some of these antennas [1–15] according to their characteristics.

1.2.1 Microstrip Antennas

The microstrip antenna is the most popular member of the printed antenna family [1–4] and is convenient to integrate with other driving circuitry of a communication system on a common PCB or a semiconductor chip.

The geometry of the microstrip antenna consists of a metal patch at the top of a dielectric substrate of a certain thickness and on the back side of the

DOI: 10.1201/9781003389859-1

substrate with an electrically large ground plane. The radiating patch may take the shape of a rectangle, circle, square, hexagon, and so on, with different dimensions so that it resonates at the operating frequency, such as $\lambda_g/2$ or $\lambda_g/4$. The most popular is the half-wave rectangular patch. The antenna may be excited using various methods such as inset feeding, proximity feeding, aperture feeding, and co-axial probing. One common approach is to feed from a microstrip line. Generally, the microstrip antenna's maximum radiation is in the broadside (perpendicular to the substrate plane) direction, and ideally there is no radiation in the end fire direction.

The relative dielectric constant of the substrate is usually in the range of $2 < \varepsilon_r < 13$. Substrates with lower ε_r have lower Q, which means higher bandwidth. The opposite is true for higher ε_r. Figure 1.1 shows a typical microstrip antenna with all of its constituent parts.

1.2.2 Slot Antennas

A slot antenna [5, 6] consists of a dielectric substrate and a narrow slit in the ground plane. It can be fed with a co-planar waveguide (CPW) transmission line, co-axial cable, slot line, waveguide, and so on. A half-wave slot antenna is mostly used due to its compact size and smooth pattern. The slot antenna is complementary to a dipole antenna, where the E-plane pattern of the slot and H-plane pattern of the dipole are omnidirectional, while the slot H-plane pattern is the same as the dipole E-plane pattern. Tapered slot antennas are used when a large bandwidth and end fire radiation are required.

The frequency range used for the application of slot antennas is 300 MHz to 30 GHz. Though slot antennas are used for navigational purposes, they have higher cross-polarization levels and low radiation efficiency. Figure 1.2 shows a slot antenna.

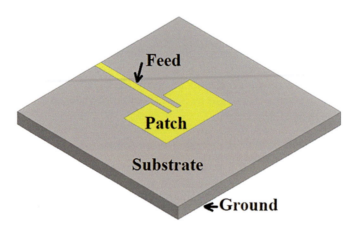

Figure 1.1 Microstrip patch antenna.

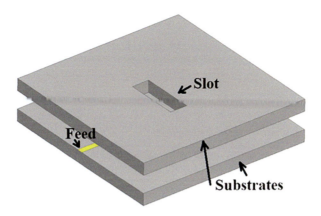

Figure 1.2 Slot antenna.

1.2.3 Dipole Antennas

The dipole antenna consists of two radiating elements fed in opposite phases that may be horizontal or vertical. A dipole antenna [7–9] can be half-wave dipole, full-wave dipole, half-wave folded dipole, or short dipole depending on its length of radiation. If the length of the dipole antenna is equal to the half wavelength, then it is half-wave dipole, and if it is equal to one wavelength, then it is a full wavelength dipole. If a two-quarter wavelength conductor is connected on both sides and folded to form a cylindrical closed shape, the feed is in the center and called a half-wave folded dipole antenna, as the total length of the dipole is half the wavelength. Again, if the length of the dipole is less than or equal to $\lambda g/10$, then it is called a short dipole. The radiation pattern of the dipole antenna is omnidirectional in the H-plane and a figure-eight in the E-plane. The most widely used dipole antenna is the half-wave dipole because of the advantage of its size, and it can operate in the frequency range of 3 KHz to 300 GHz. Figure 1.3 shows a dipole antenna.

1.2.4 Monopole Antennas

A monopole antenna consists of half of a dipole antenna. A metallic ground plane with respect to what, excitation voltage is applied to the half of the substrate, as shown in Figure 1.4. By applying image theory, the ground plane acts as a mirror; then the monopole antenna acts as a dipole antenna. The currents in monopole and dipole antennas are the same. The source voltage should be given twice in an equivalent dipole antenna compared to a monopole antenna [10–14] so that the input impedance of the monopole antenna is half of the corresponding dipole structure.

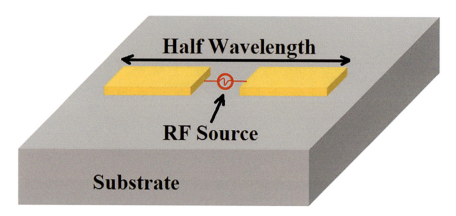

Figure 1.3 Half-wave dipole antenna.

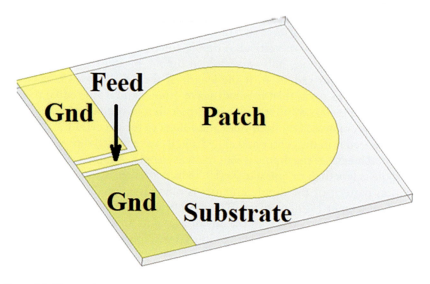

Figure 1.4 Monopole antenna.

The radiation patterns of dipole and monopole antennas are almost the same. However, the monopole antenna has a radiation field in the upper half of the space and zero radiation below the ground plane, whereas the dipole antenna has fields on both sides. The directivity of the monopole antenna is twice that of the dipole antenna. Monopole antennas are used in wideband wireless communication systems.

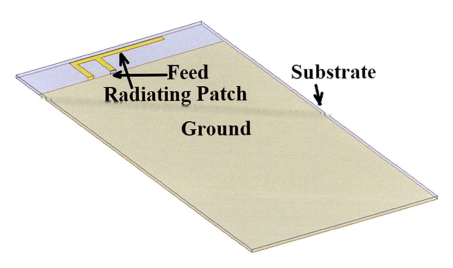

Figure 1.5 Inverted F-antennas.

1.2.5 Inverted F-Antennas

An inverted F-antenna (IFA) consists of a metallic patch and metallic ground plane. Both patch and ground plane are connected via a hole, with a metallic pin/metallic plane, as shown in Figure 1.5.

Thus, there is no dielectric substrate in it. The IFA [15, 16] antenna is actually a $\lambda g/4$ monopole antenna where the upper portion of the radiating element is folded 90° with respect to the ground plane. This bending looks like an inverted F, with one line connected to the ground, and the other line in the middle is the feed point.

The main advantage of an IFA is that it does not require any external components for impedance matching due to the hole feed arrangement, and its bandwidth can be adjusted by varying the size of the ground plane.

1.2.6 Log Periodic Antennas

This is a kind of frequency-independent antenna where the dimensions of the log periodic antenna [17, 18] are determined by the angle. A log-periodic dipole antenna (LDPA) consists of a number of half-wave dipole-driven elements of metal rods. The dipole-driven elements are connected in parallel to the feed line with alternating phase. Basically, it looks like two or more Yagi-Uda antennas connected together. Thus, adding elements of a Yagi-Uda antenna, the gain and directivity increase automatically in the log periodic antenna. Also, the LDPA has a higher bandwidth than Yagi-Uda. The one disadvantage of the LDPA is its size.

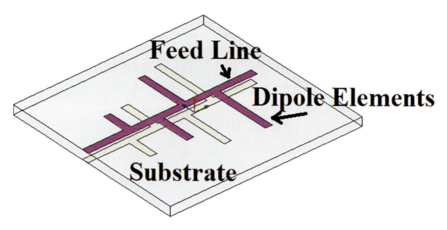

Figure 1.6 Log periodic antenna.

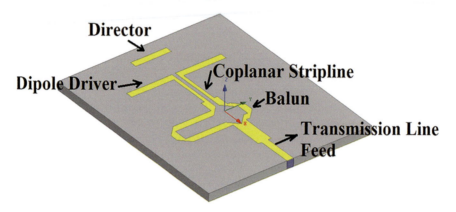

Figure 1.7 Quasi-Yagi antenna.

1.2.7 Quasi-Yagi Antennas

A planar quasi-Yagi antenna has more benefits than a wire Yagi-Uda antenna, such as a simple structure, light weight, low profile, low fabrication cost, broad bandwidth, high gain, and ease of integration with microwave circuits. The antenna radiates an endfire beam with high front-to-back ratio but narrow bandwidth. The quasi-Yagi [19, 20] antenna has improved radiation efficiency because of the truncated ground plane. The antenna works in the frequency range of 41.6 to 70.1 GHz, which is the frequency band of Q and most of the V-band. So, these types of antennas can be applied in imaging arrays and power combining.

1.2.8 Fractal Antennas

The fractal term was first implemented in 1975 by Benoit Mandelbrot as a mathematical way of understanding the infinite complexity of nature. A fractal has two unique properties: it is "self-similar" and "space filling." The idea of combining fractal geometry with electromagnetic theory in fractal antenna engineering research is gaining momentum. The self-similar property over two or more scale sizes may reduce the size of the antenna, and the property of space filling enhances bandwidth by bringing multiple resonances closer. There is an infinite number of possible geometries that can be used in designing fractal antennas, such as Koch, Minkowski, Sierpinski, Hilbert, and Cantor slot [21–24]. Figure 1.8 shows a fractal antenna.

1.2.9 Bow Tie Antennas

This antenna consists of two triangular pieces of stiff wire or two triangular flat metal plates in opposite directions, that is 180° from each other, in a bowtie-like configuration. Sometimes it is called a butterfly antenna because the flat metal plate looks like a butterfly. This antenna will show the same radiation pattern as the dipole antenna and has the linear (vertical) polarization characteristic. A bowtie antenna has better bandwidth than a dipole antenna.

This type of antenna [25] has poor transmitting efficiency in the lower frequency range. However, it is used widely for "Bolometer" applications catering the sub-mmW or THz frequency band.

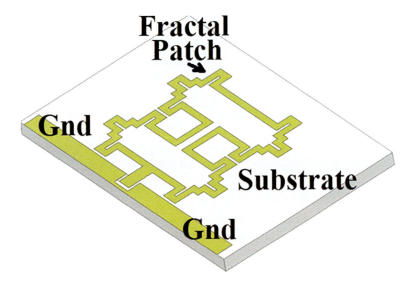

Figure 1.8 Fractal antenna.

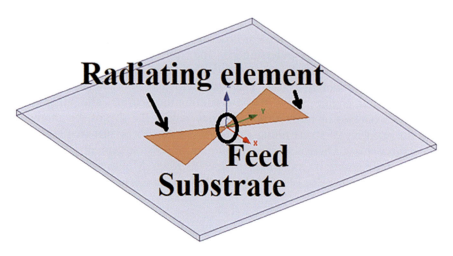

Figure 1.9 Bowtie antenna.

1.2.10 Spiral Antennas

A spiral antenna [26] consists of two conductive spirals or arms, extending from the center outwards, as shown in Figure 1.10. The direction of rotation of the spiral determines the direction of circular polarization (clockwise/anti-clockwise). Spiral antennas are one type of frequency-independent antenna whose polarization, radiation pattern, and impedance remain unchanged over large bandwidth. Generally, it is a circularly polarized antenna with low gain.

The lowest frequency of operation for a spiral antenna is approximated to occur when the wavelength is equal to the circumference of the spiral.

$$f_{low} = \frac{C}{\lambda_{low}} = \frac{C}{2\pi R_{spiral}}$$

1.2.11 Microstrip Leaky Wave Antennas

A higher-order radiative mode forms a leaky wave antenna [27, 28]. The resonant mode has odd symmetry, a null at the center, and peaks at the edges of the microstrip cross-section, so energy radiates along the long microstrip line. If the microstrip line is not terminated, then the waves reflect back from the open end, and again, a second beam will result. For long antennas, the reflected energy will reduce, and so will the second wave, so the maximum energy will radiate in the mean beam direction.

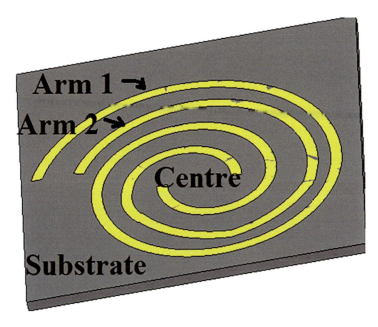

Figure 1.10 Spiral antenna.

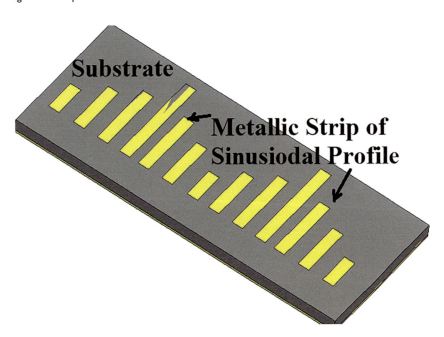

Figure 1.11 Microstrip leaky wave antenna.

The most significant characteristic of microstrip leaky wave antennas is their beam scanning capacity.

1.3 SUMMARY

This chapter outlines various types of printed antennas used in day-to-day activities. The basics of those antenna categories have been explained, with their end uses. In practice, there are numerous antenna types, but here, only major categories have been discussed. This gives an overview of the practical applications along with three-dimensional schematics. In the next chapter, we will see how an effective method is adopted in modern antenna engineering for miniaturization purposes.

REFERENCES

[1] R. B. Waterhouse, *Microstrip Patch Antennas: A Designer's Guide*, Norwell, MA: Kluwer Academic Publishers, 2003.

[2] Y. Al-Naiemy, *Design, Fabrication, and Testing of Microstrip Antennas*, London: Lambert Academic Publishing, 2013.

[3] B. KI. Pandey, A. K. Pandey, S. B. Chakrabarty, S. Kulshrestha, A. L. Solanki, and S. B. Sharma, "Dual Frequency Single Aperture Microstrip Patch Antenna Element for SAR Applications," *IETE Technical Review*, Vol. 23, no. 6, pp. 357–366, 2006.

[4] C. Mendes, and C. Peixeiro, "A Dual-Mode Single-Band Wearable Microstrip Antenna for Body Area Networks," *IEEE Antennas and Wireless Propagation Letters*, Vol. 16, pp. 3055–3058, 2017.

[5] M. A. Antoniades, A. Dadgarpour, A. R. Razali, A. Abbosh, and T. A. Denidni, "Planar Antennas for Compact Multiband Transceivers Using a Microstrip Feedline and Multiple Open-Ended Ground Slots," *IET Microwaves, Antennas & Propagation*, Vol. 9, no. 5, pp. 486–494, 2015.

[6] H. Liu, A. Qing, B. Wu, and Z. Xu, "Low-Profile Hexagonal SIW Cavity Slot Antenna with Enhanced Gain by Using Quasi-TM310 Mode," *IET Microwaves, Antennas & Propagation*, Vol. 14, no. 7, pp. 629–633, 2020, doi: 10.1049/iet-map.2019.0626.

[7] Z. Zhou, Z. Wei, Z. Tang, and Y. Yin, "Design and Analysis of a Wideband Multiple-Microstrip Dipole Antenna with High Isolation," *IEEE Antennas and Wireless Propagation Letters*, Vol. 18, no. 4, pp. 722–726, 2019, doi: 10.1109/LAWP.2019.2901838.

[8] S. Yan, P. J. Soh, and G. A. E. Vandenbosch, "Wearable Dual-Band Magneto-Electric Dipole Antenna for WBAN/WLAN Applications," *IEEE Transactions on Antennas and Propagation*, Vol. 63, no. 9, pp. 4165–4169, 2015, doi: 10.1109/TAP.2015.2443863.

[9] L. Ge, and K. M. Luk, "A Low-Profile Magneto-Electric Dipole Antenna," *IEEE Transactions on Antennas and Propagation*, Vol. 60, no. 4, pp. 1684–1689, 2012, doi: 10.1109/TAP.2012.2186260.

[10] C. Deng, Y. Xie, and P. Li, "CPW-Fed Planar Printed Monopole Antenna with Impedance Bandwidth Enhanced," *IEEE Antennas and Wireless Propagation Letters*, Vol. 8, pp. 1394–1397, 2009, doi: 10.1109/LAWP.2009.2039743.

[11] M. A. Al-Mihrab, A. J. Salim, and J. K. Ali, "A Compact Multiband Printed Monopole Antenna With Hybrid Polarization Radiation for GPS, LTE, and Satellite Applications," *IEEE Access*, Vol. 8, pp. 110371–110380, 2020, doi: 10.1109/ACCESS.2020.3000436.

[12] H. Huang, Y. Liu, S. Zhang, and S. Gong, "Multiband Metamaterial-Loaded Monopole Antenna for WLAN/WiMAX Applications," *IEEE Antennas and Wireless Propagation Letters*, Vol. 14, pp. 662–665, 2015, doi: 10.1109/ LAWP.2014.2376969.

[13] N. P. Agrawall, G. Kumar, and K. P. Ray, "Wide-Band Planar Monopole Antennas," *IEEE Transactions on Antennas and Propagation*, Vol. 46, no. 2, pp. 294–295, 1998, doi: 10.1109/8.660976.

[14] K. Yen-Liang, and W. Kin-Lu, "Printed Double-T Monopole Antenna for 2.4/5.2 GHz Dual-Band WLAN Operations," *IEEE Transactions on Antennas and Propagation*, Vol. 51, no. 9, pp. 2187–2192, 2003, doi: 10.1109/ TAP.2003.816391.

[15] X. Yang, H. Wong, and J. Xiang, "Polarization Reconfigurable Planar Inverted-F Antenna for Implantable Telemetry Applications," *IEEE Access*, Vol. 7, pp. 141900–141909, 2019, doi: 10.1109/ACCESS.2019.2941388.

[16] T. Houret, L. Lizzi, F. Ferrero, C. Danchesi, and S. Boudaud, "DTC-Enabled Frequency-Tunable Inverted-F Antenna for IoT Applications," *IEEE Antennas and Wireless Propagation Letters*, Vol. 19, no. 2, pp. 307–311, 2020, doi: 10.1109/LAWP.2019.2961114.

[17] S. Kurokawa, M. Hirose, and K. Komiyama, "Measurement and Uncertainty Analysis of Free-Space Antenna Factors of a Log-Periodic Antenna Using Time-Domain Techniques," *IEEE Transactions on Instrumentation and Measurement*, Vol. 58, no. 4, pp. 1120–1125, 2009, doi: 10.1109/TIM.2008.2012383.

[18] C. A. Balanis, "Frequency-Independent Antennas: Spirals and Log-Periodics," *Modern Antenna Handbook* (Wiley), pp. 263–323, 2008, doi: 10.1002/9780470294154.ch6.

[19] P. Qin, A. R. Weily, Y. J. Guo, T. S. Bird, and C. Liang, "Frequency Reconfigurable Quasi-Yagi Folded Dipole Antenna," *IEEE Transactions on Antennas and Propagation*, Vol. 58, no. 8, pp. 2742–2747, 2010, doi: 10.1109/ TAP.2010.2050455.

[20] N. Kaneda, W. R. Deal, Yongxi Qian, R. Waterhouse, and T. Itoh, "A Broadband Planar Quasi-Yagi Antenna," *IEEE Transactions on Antennas and Propagation*, Vol. 50, no. 8, pp. 1158–1160, 2002, doi: 10.1109/TAP.2002. 801299.

[21] Z. Yu, J. Yu, C. Zhu, and Z. Yang, "An Improved Koch Snowflake Fractal Broadband Antenna for Wireless Applications," *IEEE 5th International Symposium on Electromagnetic Compatibility* (EMC-Beijing), pp. 1–5, 2017, doi: 10.1109/EMC-B.2017.8260462.

[22] M. Comisso, "Theoretical and Numerical Analysis of the Resonant Behaviour of the Minkowski Fractal Dipole Antenna," *IET Microwaves, Antennas & Propagation*, Vol. 3, no. 3, pp. 456–464, 2009, doi: 10.1049/ iet-map.2008.0249.

[23] K. C. Hwang, "A Modified Sierpinski Fractal Antenna for Multiband Application," *IEEE Antennas and Wireless Propagation Letters*, Vol. 6, pp. 357–360, 2007, doi: 10.1109/LAWP.2007.902045.

[24] J. Li, T. Jiang, C. Wang, and C. Cheng, "Optimization of UHF Hilbert Antenna for Partial Discharge Detection of Transformers," *IEEE Transactions on Antennas and Propagation*, Vol. 60, no. 5, pp. 2536–2540, 2012, doi: 10.1109/TAP.2012.2189929.

[25] T. Li, H. Zhai, X. Wang, L. Li, and C. Liang, "Frequency-Reconfigurable Bow-Tie Antenna for Bluetooth, WiMAX, and WLAN Applications," *IEEE Antennas and Wireless Propagation Letters*, Vol. 14, pp. 171–174, 2015, doi: 10.1109/LAWP.2014.2359199.

[26] Y. Han, K. Hu, R. Zhao, Y. Gao, L. Dai, Y. Fu, B. Zhou, and S. Yuan, "Design of Combined Printed Helical Spiral Antenna and Helical Inverted-F Antenna for Unmanned Aerial Vehicle Application," *IEEE Access*, Vol. 8, pp. 54115–54124, 2020, doi: 10.1109/ACCESS.2020.2981041.

[27] Y. Li, Q. Xue, E. K. Yung, and Y. Long, "The Periodic Half-Width Microstrip Leaky-Wave Antenna With a Backward to Forward Scanning Capability," *IEEE Transactions on Antennas and Propagation*, Vol. 58, no. 3, pp. 963–966, 2010, doi: 10.1109/TAP.2009.2039304.

[28] D. R. Jackson, and A. A. Oliner, "A Leaky-Wave Analysis of the High-Gain Printed Antenna Configuration," *IEEE Transactions on Antennas and Propagation*, Vol. 36, no. 7, pp. 905–910, 1988, doi: 10.1109/8.7194.

Problems

1. Why are printed versions of antennas needed?
2. Why does an antenna act as a transducer?
3. How is a monopole antenna different from a dipole antenna?
4. Name a few frequency-independent antennas.
5. What are broad-side and end-fire arrays?
6. What is the physical significance of substrate height in antennas?
7. Where is the bowtie antenna used extensively?
8. What are the different types of feeding techniques implemented in printed antenna?
9. Which feeding technique is most efficient?
10. What are complementary antenna types? Give examples.
11. What is a MIMO antenna? Why is it becoming popular?
12. What is a GPS antenna?
13. What frequency range is used for 4G-LTE communication? What about 5G?
14. What are the G/T ratio and EIRP of antennas?
15. Why are circularly polarized antennas usually preferred?
16. What are balanced and unbalanced transmission lines?

17. What are the calibration methods used for antenna VSWR measurement? Why are they required?
18. How can you increase the power handling capability of printed antennas?
19. How do you overcome the various disadvantages of microstrip antennas?
20. What are the different gain enhancement techniques in printed antennas?

Chapter 2

Fractals in Printed Antennas

2.1 INTRODUCTION

In recent years, there has been a high demand for light weight, compact, miniaturized, and multiband antennas for wireless communication and their integration into smartphones, palmtops, cars, airplanes, and many other embedded systems. So, the use of ultra-wideband antennas is necessary. Designing such antennas with a fractal structure gives multiband and broadband behavior. A fractal can be defined as a fragmented geometric shape with a fractional dimension that can be subdivided into parts; each part is an approximate reduced copy of the whole.

The original inspiration for the development of fractal geometry came largely from patterns found in nature. An interesting fact about fractals is that they are the best existing mathematical expression of many natural forms, such as coastlines, mountains, or parts of living organisms. The properties of fractals are extremely irregular or fragmented elements. The leading researcher in the field of fractals, Benoit Mandelbrot, developed fractals in 1975 [1, 2] as a mathematical way of understanding the infinite complexity of nature. This is called fractal geometry. Fractal geometry [3] offers a very simple process of iteration, completely different from classical geometry, which uses formulas to define a shape. Fractals have two unique properties; they are self similar and space filling.

The key aspect lies in the repetition or self similar property over two or more scale sizes that may reduce the size of antenna. The property of space filling enhances bandwidth by bringing multiple resonances closer. That is why, it is capable of giving well to excellent performance at many different frequencies simultaneously. This way, fractal engineering is welcomed in antenna field widely. Fractal antennas are very compact in size and excellent in design for wideband and multiband applications. There has been a possibility of designing new types of antennas using fractal geometry rather than Euclidean geometry. The idea of combining fractal geometry with electromagnetic theory in fractal antenna engineering research [4] is thus gaining momentum.

DOI: 10.1201/9781003389859-2

Here a preliminary introduction to different fractal antennas is given, and different UWB antennas are also introduced based on different fractal shapes on the patch/ground plane. Figures 2.1–2.7 show different popular fractal geometries.

2.2 CANTOR SET

The Cantor set was invented by the German mathematician Georg Cantor. It is built iteratively from a segment by removing a central portion; then the operation is repeated on the remaining two segments, and so on.

2.3 KOCH CURVE

Begin with a straight line, divide it into three equal segments, and replace the middle segment with the two sides of an equilateral triangle of the same length as the segment being removed (the two segments in the middle figure).

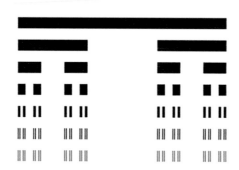

Figure 2.1 Cantor set.

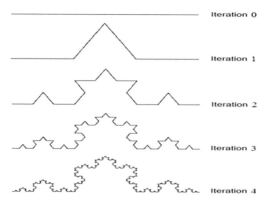

Figure 2.2 Koch curve.

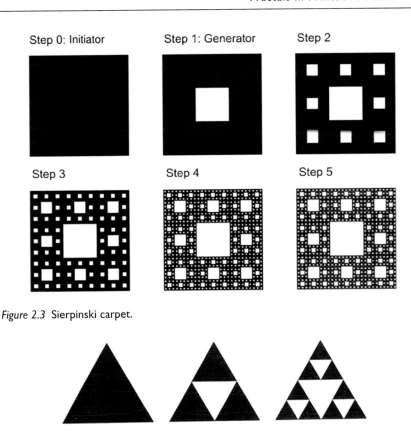

Figure 2.3 Sierpinski carpet.

Figure 2.4 Sierpinski gasket.

Now repeat, taking each of the four resulting segments, dividing them into three equal parts and replacing each of the middle segments by two sides of an equilateral triangle. Continue this construction.

2.4 SIERPINSKI CARPET

The Sierpinski carpet is a plane fractal first described by Wacław Sierpiński in 1916. The carpet is one generalization of the Cantor set to two dimensions; another is Cantor dust.

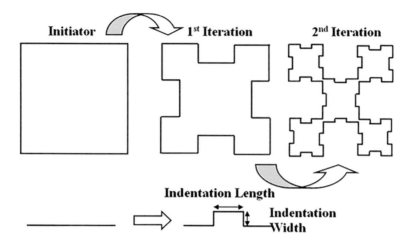

Figure 2.5 Minkowski fractal.

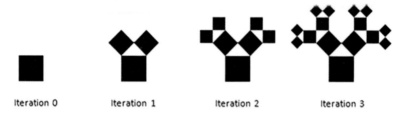

Figure 2.6 Pythagoras tree fractal.

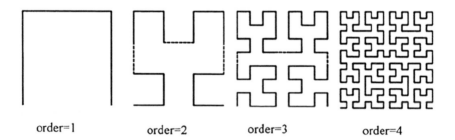

Figure 2.7 Hilbert curve fractal.

2.5 SIERPINSKI GASKET

The Sierpinski triangle (also with the original orthography, Sierpiński), also called the Sierpinski gasket or the Sierpinski sieve, is an attractive fractal fixed set with the overall shape of an equilateral triangle, subdivided recursively into smaller equilateral triangles.

2.6 MINKOWSKI FRACTAL

The Minkowski content (named after Hermann Minkowski), or the boundary measure, of a set is a basic concept that uses concepts from geometry and measure theory to generalize the notions of length of a smooth curve in the plane and an area of a smooth surface in space to arbitrary measurable sets.

2.7 PYTHAGORAS TREE FRACTAL

The Pythagoras tree is a plane fractal constructed from squares. Invented by the Dutch mathematics teacher Albert E. Bosman in 1942, it is named after the ancient Greek mathematician Pythagoras because each triple of touching squares encloses a right triangle in a configuration traditionally used to depict the Pythagorean theorem. The construction of the Pythagoras tree begins with a square. Upon this square are constructed two squares, each scaled down by a linear factor of $\sqrt{2}/2$, such that the corners of the squares coincide pairwise. The same procedure is then applied recursively to the two smaller squares ad infinitum. Figure 2.6 shows the first small iteration in the construction process.

2.8 HILBERT CURVE FRACTAL

A Hilbert curve (also known as a Hilbert space-filling curve) is a continuous fractal space-filling curve first described by the German mathematician David Hilbert in 1891 as a variant of the space-filling Peano curves discovered by Giuseppe Peano in 1890.

Because it is space-filling, its Hausdorff dimension is 2 (precisely, its image is the unit square, whose dimension is 2 in any definition of dimension; its graph is a compact set homeomorphic to the closed unit interval, with Hausdorff dimension 2).

2.9 BACKGROUND STUDY OF FRACTALS IN ANTENNA ENGINEERING

Fractal antennas have useful applications in cellular telephone and microwave communications. Video conferencing and streaming video are main applications that are included in next-generation networks, and requirements for these applications are high data rates with high bandwidth. But as the size of the antenna decreases, bandwidth support also decreases. So it is required to have a small size with high bandwidth. Applying fractals to antenna elements allows for smaller resonant antennas that are multiband/broadband and may be optimized for gain.

2.9.1 UWB Fractal Antennas With Elliptical/Circular Patches

Figure 2.8 shows a fractal antenna, which is based on an elliptical patch. The height of the initial elliptical monopole antenna is chosen as $\lambda g/4$ with the lowest resonant frequency. By etching a Sierpinski fractal curve slot of the third iteration, the lowest frequency of the operating band shifts from 3 to 2.8 GHz. The resonance [5] characteristics at lower frequencies are also improved. Also, the impedance matching at higher frequencies is improved by embedding a Von Koch fractal on each side of the hexagonal boundary of the planar quasi-self-complementary antenna [6]. The gain variation of this type of UWB antenna is acceptable for indoor wireless communication over short ranges.

The first iteration of a crown-shaped fractal antenna is constructed by inscribing an equilateral triangle patch inside the circle and subtracting it from the circle [7]. The second iteration is achieved by constructing an inner circle and an inscribed equilateral triangle, subtracted from that inner circle. Likewise, other iterations are developed as shown in Figure 2.9(a). This achieves impedance bandwidth larger than 6:1. Fattahi et al. [8] follow the previous iteration rules by inscribing hexagonal types of patches inside original circular patch. Dorostkar et al. [9] designed a circular-hexagonal fractal with super-wide bandwidth ranging from 2.18–44.5 GHz with a bandwidth ratio of 20.4:1. The bandwidth ratio is much better than that of recently reported super-wideband antennas, which makes it appropriate for many wireless communications systems such as ISM, Wi-Fi, GPS, Bluetooth, WLAN, and UWB.

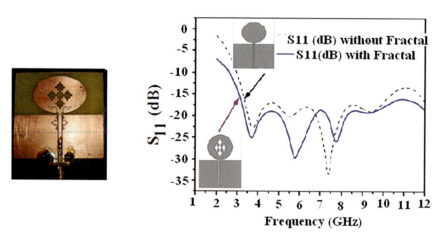

Figure 2.8 (a) Fractal monopole UWB antenna with elliptical patch. (b) S11 characteristics of the antenna with and without fractal [5].

Source: © Copyright JEMWA

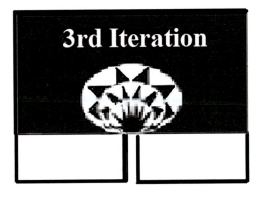 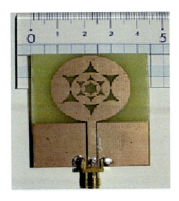

Figure 2.9 Fractal monopole UWB antennas with circular patch [7, 14].
Source: © Copyright MOTL

Mehetra et al. [10] designed different types of fractal antennas. The Apollo fractal structure of an antenna, as described by Mehetra et al., is formed by adjusting three circles of similar diameter in such a way that all three circles are in close touch with each other. The triangular wheel-shaped fractal [11] exhibits excellent bandwidth from 0.86 to 11 GHz. The fourth iterated antenna shifts the lower-band edge frequency from 1.44 to 0.86 GHz in comparison to a simple CPW-fed circular patch with the same dimensions. A third iterated tree-type fractal antenna is reported [12] that exhibits a bandwidth of 5 to 14 GHz, and a third iterated triangular non-concentric circular fractal antenna exhibits 1.335 to 14.8 GHz bandwidth [13]. A flower-shaped fractal antenna, as shown in Figure 2.9(b), has been proposed [14], which corresponds to 138% fractional bandwidth. A novel smiley fractal antenna (SFA) [15], employed with N-notch feed and modified ground plane, is designed and developed to achieve the desired characteristics.

2.9.2 UWB Fractal Antennas With Square/Rectangular Patch

Different types of UWB fractal antennas based on square or rectangular patch have been reported. A Sierpinski carpet fractal antenna is developed [16] with 4.65–10.5 GHz (VSWR < 2) impedance bandwidth. This antenna has been proposed on a grounded CPW to improve antenna behavior. This new design-fed technique changes the behavior of the fractal elements from multiband to wideband. The fractal concept has been used for widening the bandwidth of printed monopole antennas to comply with UWB requirements. The concept of self-similarity has been deployed and shows this type of antenna where repeated hexagonal shapes form composite monopoles to increase their frequency range. The repeated shapes act as multi-resonance elements and thus provide wider bandwidth. A Sierpinski carpet fractal is thus formed on the first iteration of the Minkowski fractal. The proposed antenna is miniaturized (45 × 45 mm) and has multiband characteristics. Simulation results show that the presented antenna has a reflection coefficient of < −15 dB. Islam et al. [17] presented a printed micro-strip line-fed fractal antenna with a high fidelity factor for UWB microwave imaging applications, as shown in Figure 2.10. The fractal antenna achieves fractional bandwidth of 117.88%, covering the frequency range from 3.1 to more than 12 GHz with a maximum gain of 3.54 dBi at 6.25 GHz. A compact fractal antenna is fabricated that gives satisfactory results in the UWB spectrum with the impedance bandwidth ranging from 3.4 to 14.5 GHz with a fractional bandwidth of 124.03% and average gain of 4 dBi [18].

2.9.3 UWB Fractal Antennas of Different Shapes

Different types of fractal antenna are illustrated in Figure 2.11(a–f) [19–29]. Figure 2.11 (a) to (e) shows five different types of tree-shaped fractal antenna. The second tree-type antenna is made up of a radiating patch composed of a repeating pattern of unit cells [19] that creates a fractal tree; enhancement of the antenna's bandwidth is achieved by increasing the unit cells of the fractal tree without significantly affecting the antenna's physical size. A novel modification of a Pythagorean [20] fractal tree monopole antenna has been implemented just by increasing the tree fractal iterations. New resonances are obtained and operate over the frequency band between 2.6 and 11.12 GHz for VSWR < 2. In Figure 2.11(d and e), the proposed fractal antennas [21, 22] exhibit wider impedance bandwidth from 2 to 13.7 GHz and 4.3 to 15.5 GHz, respectively, for VSWR > 2. By increasing the number of fractal iterations, the impedance bandwidth of both fractal antennas is increased. Jalali and Sedghi [23] designed a miniaturized fractal antenna composed of fractal rectangular unit cells. The antenna has dimensions of 14 × 18 × 1 mm^3

Figure 2.10 Different types of fractal monopole UWB antennas with square/rectangular patch. [17]

Source: ©Copyright MOTL

Fractals in Printed Antennas 23

Figure 2.11 Different types of fractal UWB monopole antennas [19, 21–22, 24–25, 27].
Source: © Copyright MOTL

over a frequency band from 2.95 to 12.81 GHz, with a fractional bandwidth of 125%. Again, a Minkowski-like fractal geometry-based ultra-wideband monopole antenna [24] is proposed [Figure 2.11(d)]. A new type of fractal is designed [Figure 2.11(e)] that covers a 3.15- to 19-GHz frequency band [25], and a hexagonal fractal with Koch geometry [26] and Sierpinski fractal [27] are used for bandwidth enhancement with 122% and 2.4 to 12.1 GHz, respectively. The composition of co-planar waveguide feeding with conventional standing fractal monopole antenna using Penta-Gasket-Koch (PGK) is introduced [28]. For better insight, this antenna is compared with a simple feed using the same radiation element. The return loss of the main design achieves a good input impedance match and linear phase of S_{11} throughout the pass band (4.25–11 GHz).

The challenge of designing compact antennas with operating bandwidths of more than 100% can be met by using wide-slot antennas with fractal boundaries to increase the antenna's electrical length for operation at lower-frequency bands. Again, it has been found that with the addition of small fractal elements at the corners of a polygon patch, standard UWB bandwidth can be covered [29].

2.9.4 UWB Fractal Antennas With Ground Plane

Chen et al. [30] designed a UWB antenna by implementing fractal geometry in the ground plane. In Figure 2.12(a), a micro-strip–fed planar antenna is embedded in a semi-elliptically fractal-complementary slot onto the

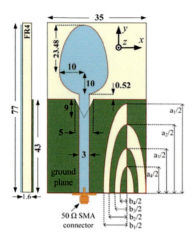 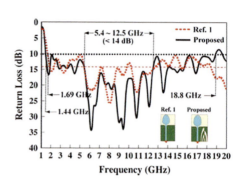

Figure 2.12 (a) UWB fractal antenna based on ground plane (b) S_{11} characteristic of proposed antenna with comparison of ref antenna [30].

Source: © Copyright IEEE

asymmetrical ground plane; a 10-dB bandwidth of 172% (1.44–18.8 GHz) is thus achieved with a bandwidth > 12:1 ratio. The semi-elliptically complementary ground fractal reduces lower-band edge frequency from 1.69 to 1.44, GHz as shown in Figure 2.12(b).

2.10 RECENT WORKS ON FRACTALS IN ANTENNA ENGINEERING

Initially, a circular monopole antenna based on Descartes circle theorem was designed for UWB application. By implementing Apollonian fractal geometry [31] in the radiating patch, the antenna achieved 142% (1.8–10.6 GHz) fractional bandwidth with VSWR < 2. Initially the dimension of circular radiation was taken by $\lambda_g/2$ of the lowest resonant frequency of 2 GHz. Using fractal geometry up to the third iteration, a self-similar iterative structure was constructed, as shown in Figure 2.13(a) [32]. In the first iteration, three circles were etched from a circular radiator, each with a radius of 6.3 mm, which was found by dividing the original circle by $\left(\frac{3+2\sqrt{3}}{3}\right)$. In the second iteration, similarly, three circles were etched with radius 3.5 mm by dividing the original circle by $(1+2\sqrt{3})$. In the third iteration, ten circles were subtracted from the radiator by dividing the original circle by $\left(5+\frac{8}{\sqrt{3}}\right)$.

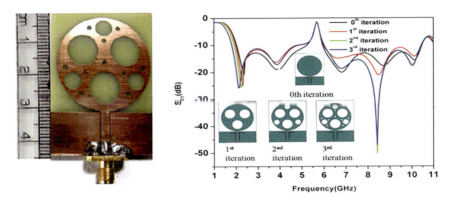

Figure 2.13 (a) Prototype antenna. (b) Step-by-step iteration and response.

The fractal geometry has two properties: it is self similar and space filling. The self-similar property brings higher-order modes close to each other and thus gives multiband characteristics, and the space-filling property tends to fill the area occupied by the antenna as the order of iteration increases and thus is useful for minimizing the size of the antenna. Figure 2.13(b) shows the steps for miniaturization of a planar antenna. Also, a pair of L-shaped slots is inserted for notching out the frequency of the 5.5-GHz WLAN band. Due to fabrication constrains; the design is still third iteration. It can be seen that there is a shift to a lower frequency for S_{11} better than −10 dB as iteration increases [please refer to Figure 2.13(b)]. The introduction of the fractal shape enhances the effective electrical path of the surface current, which in turn increases the effective impedance bandwidth.

We know that the resonance characteristics of a coplanar waveguide–based UWB monopole antenna also depend on the shape and size of the ground plane. Now another miniaturization technique is carried out by the implementation of fractal geometry in the ground plane, as shown in Figure 2.14. This technique gives a fractional impedance bandwidth of 114% (3.1–11.3 GHz) and reduces the overall dimensions of the antenna. A modified crown square fractal is created in the CPW ground plane by etching a rectangle and thereafter adding a square from the middle portion of the rectangle in an iterative manner. The self-similar and space-filling properties of fractal geometry increase the effective electrical length to reduce the size of the antenna and enhance bandwidth by bringing multiple resonances closer. Thus, the overall bandwidth increases by 58% in comparison to the simple square monopole antenna. Figure 2.15 [33] gives the details of the antenna miniaturization technique. Two omega-shaped slots are inserted at the radiating square patch for notching out a 5.5-GHz WLAN band and 7.5-GHz satellite communication frequency band.

Figure 2.14 Prototype antenna.

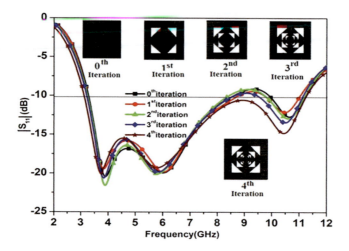

Figure 2.15 Step-by-step iteration and response.

The radiation patterns of the proposed antenna resemble a conventional monopole at lower frequencies, which is nearly omnidirectional in the H-plane and a figure-eight in the E-plane, as depicted in Figure 2.16.

Modern communications demand compact, lighter-weight antennas compatible with portable devices for multiband utilization. To fulfill this requirement, a miniaturized planar monopole antenna is designed for multi-band utilization [34–35] that satisfies the performance requirements of both UWB and Bluetooth technologies with this single antenna. Inserting a second iterated fractal binary tree in a modified half annular-shaped antenna [36] increased the functionality with an extra Bluetooth band at 2.45 GHz

Fractals in Printed Antennas 27

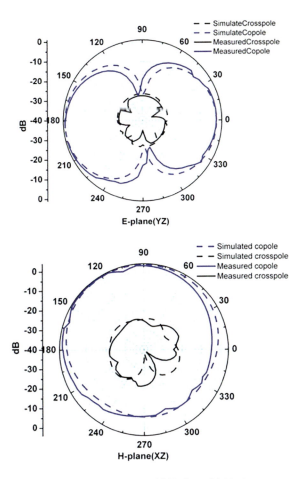

Figure 2.16 Radiation pattern of the antenna: (a) E-plane (b) H-plane.

without affecting the performance of the UWB band. The tree geometry starts with a stem allowing one of its ends to branch off in two directions. In the next stage of iteration, each of these branches is allowed to branch off again. The fractal concept for the Bluetooth band reduces the overall antenna size. The antenna and its S_{11} characteristics are shown in Figure 2.17(a and b), respectively.

Recently, the use of radio frequencies in microwave imaging has received remarkable impetus in diagnosing tumors and detection of malignant tissues in the human body. Miniaturized antennas with highly directional patterns and enhanced gain characteristics are essential for this particular application. Vivaldi antennas have served this purpose due to their inherent broadband nature, along with their high directivity and gain.

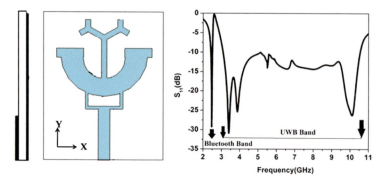

Figure 2.17 (a) Proposed antenna. (b) S_{11} characteristics of the antenna.

Figure 2.18 Prototype antenna. (a) Front view. (b) Back view.

A new approach to designing a compact, high-gain antipodal Vivaldi antenna is implemented by introducing a fern leaf–type fractal structure, as shown in Figure 2.18 [37]. The second iterated fern fractal structure results in a fractional bandwidth of 175.58% (1.3 to 20 GHz), as shown in Figure 2.19, and provides a 60% size reduction compared to the standard antenna, as shown in Table 2.1.

The correlation property of the pulse signals is determined by the fidelity factor, which describes how the received signal varies in different orientations with respect to the signal in the main beam direction, as shown in Figure 2.20. The fidelity factor between the signal at the main beam direction and the signal in a particular angular direction is defined as the normalized cross-correlation between them. It is found that the overall fidelity factors are much better in the E-plane than the H-plane. It has been observed that, up to a 60° angle of rotation, the quality of fidelity factors is acceptable; beyond that, they starts to degrade. The reason behind it is, the phase

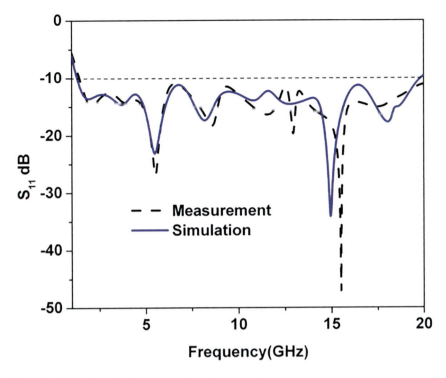

Figure 2.19 Simulated and measured S_{11} characteristics.

Table 2.1 Comparison of Proposed Antenna With Traditional Antipodal Vivaldi Antenna

Antenna	Size (mm³)	Gain (dBi)	Efficiency (%)	Complexity
Traditional anti-podal Vivaldi	(90 × 87 × 0.8)/4.4 = 1423.6	7	84%	Easy to design
Proposed antenna	(50.8 × 62 × 0.8)/4.4 = 572.6	8	96%	Complex

difference of transmitted and received signals approaches to 90°. Thus, the two signals become orthogonal to each other, which results in lowering the value of the fidelity factor, ideally approaching zero.

2.11 SUMMARY

In this chapter, some of the most popular fractal geometrics are presented, like Koch, Sierpinski fractal, Minkowski fractal, and Hilbert fractal. The common feature of all the presented works is that fractal antennas allow

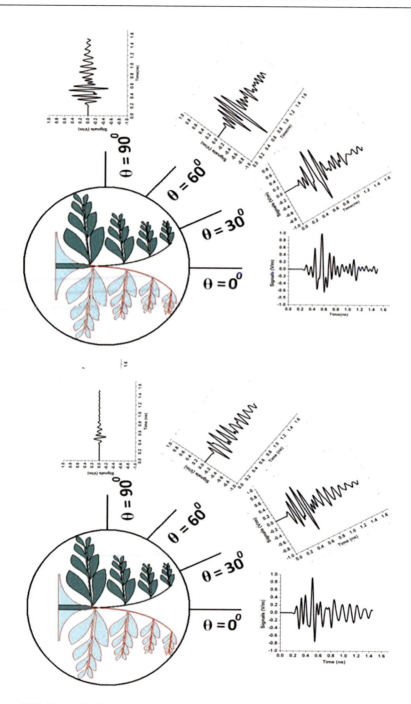

Figure 2.20 Normalized radiation signal waveform in (a) E-plane and (b) H-plane.

broadband and a multiband behavior. In the literature survey, it was observed that fractals are used successfully in antenna engineering and give satisfactory performances in the design of miniaturized antennas. There has been a possibility of designing new types of antennas using fractal geometry rather than Euclidean geometry. Great efforts are still going on to develop new UWB miniaturized antennas with the desired performance. Still more studies are needed for their proper exploitation.

REFERENCES

[1] B. B. Mandelbrot, *Fractals and Chaos: The Mandelbrot Set and Beyond*, Berlin: Springer, ISBN 978-0-387-20158-0, 2004.

[2] B. B. Mandelbrot, *The Fractal Geometry of Nature*. San Francisco, CA: Freeman, 1983.

[3] F. Kenneth, *Fractal Geometry: Mathematical Foundations and Applications*. Hoboken: John Wiley & Sons, p. xxv, ISBN 0-470-84862-6, 2003.

[4] H. D. Werner, and S. Ganguly, "An Overview of Fractal Antenna Engineering Research," *IEEE Antennas and Propagation Magazine*, Vol. 45, no. 1, pp. 38–57, 2003.

[5] A. Gorai, A. Karmakar, M. Pal, and R. Ghatak, "Multiple Fractal-Shaped Slots-Based UWB Antenna with Triple-Band Notch Functionality," *Journal of Electromagnetic Waves and Applications*, Vol. 27, no. 18, pp. 2407–2415, 2013.

[6] A. Gorai, M. Pal, and R. Ghatak, "A Compact Fractal-Shaped Antenna for Ultrawide Band and Bluetooth Wireless Systems with WLAN Rejection Functionality," *IEEE Antennas and Wireless Propagation Letters*, Vol. 16, pp. 2163–2166, 2017.

[7] R. Kumar, and K. Sawant, "On the Design of Inscribed Triangle Non-Concentric Circular Fractal Antenna," *Microwave and Optical Technology Letters*, Vol. 52, no. 12, pp. 2696–2699, 2010.

[8] J. Fattahi, C. Ghobadi, J. Nourinia, and D. Ahmadian, "A New CPW-Fed Slot Antenna Using Fractal Technique for Ultra Wide Band Application," *International Journal of Microwave and Optical Technology*, Vol. 7, no. 2, pp. 2846–2850, 2012.

[9] M. A. Dorostkar, M. T. Islam, and R. Azim, "Design of a Novel Super Wide Band Circular Hexagonal Fractal Antenna," *Progress in Electromagnetics Research C*, Vol. 139, pp. 229–245, 2013.

[10] T. R. Mehetra, and R. Kumar, "Design of Inscribed Circle Apollo UWB Fractal Antenna with Modified Ground Plane," *Indian Journal of Science and Technology*, Vol. 5, no. 6, pp. 116–120, 2012.

[11] R. Kumar, and P. Malathi, "On the Design of CPW-Fed Ultra Wide Band Triangular Wheel Shape Fractal Antenna," *International Journal of Microwave and Optical Technology*, Vol. 5, no. 2, pp. 89–93, 2010.

[12] R. Kumar, and B. N. Chaubey, "On the Design of Tree Type Ultra Wide Band Fractal Antenna for DS-CDMA System," *Journal of Microwaves, Optoelectronics and Electromagnetic Applications*, Vol. 11, no. 1, pp. 107–121, 2012.

[13] R. Kumar, and K. Sawant, "On the Design of Inscribed Triangular Non Centric Circular Fractal Antenna," *Microwave and Optical Technology Letters*, Vol. 52, no. 12, pp. 2696–2699, 2010.

[14] D. J. Kim, J. H. Choi, and Y. S. Kim, "CPW-Fed Ultra Wideband Flower Shaped Circular Fractal Antenna," *Microwave and Optical Technology Letters*, Vol. 55, no. 8, pp. 1792–1796, 2013.

[15] M. Susila, T. R. Rao, and A. Gupta, "A Novel Smiley Fractal Antenna (SFA) Design and Development for UWB Wireless Applications," *Progress in Electromagnetics Research C*, Vol. 54, pp. 171–178, 2014.

[16] M. N. Jahromi, A. Falahati, and R. M. Edwards, "Bandwidth and Impedance Matching Enhancement of Fractal Monopole Antennas Using Compact Grounded Coplanar Waveguide," *IEEE Transactions on Antennas and Propagation*, Vol. 59, no. 7, pp. 2480–2487, 2011.

[17] M. M. Islam, M. T. Islam, M. Samsuzzaman, M. R. I. Faruque, and N. Misran, "Microstrip Line-Fed Fractal Antenna with a High Fidelity Factor for UWB Imaging Applications," *Microwave and Optical Technology Letters*, Vol. 57, no. 11, pp. 2580–2585, 2015.

[18] A. Pandey, S. Singhal, and A. K. Singh, "CPW-Fed Third Iterative Square-Shaped Fractal Antenna for UWB Application," *Microwave and Optical Technology Letters*, Vol. 58, no. 1, pp. 92–99, 2016.

[19] M. N. Moghadesi, R. A. Sadeghzadeh, T. Aribi, T. Sedghi, and B. S. Virdee, "UWB Monopole Microstrip Antenna Using Fractal Tree Unit-Cell," *Microwave and Optical Technology Letters*, Vol. 54, no. 10, pp. 2366–2370, 2012.

[20] J. Pourahmadazar, C. Ghobadi, and J. Nourinia, "Novel modified Pythagorean Tree Fractal Monopole Antennas for UWB Application," *IEEE Antennas and Wireless Propagation Letters*, Vol. 10, pp. 484–487, 2011.

[21] V. Rajeshkumar, and A. Raghavan, "Bandwidth Enhanced Compact Fractal Antenna for UWB Applications with 5–6 GHz Band Rejection," *Microwave and Optical Technology Letters*, Vol. 57, no. 3, pp. 607–613, 2015.

[22] S. Singhal, T. Goel, and A. K. Singh, "Inner Tapered Tree-Shaped Fractal Antenna for UWB Applications," *Microwave and Optical Technology Letters*, Vol. 57, no. 3, pp. 559–567, 2015.

[23] M. Jalali, and T. Sedghi, "Very Compact UWB CPW-Fed Fractal Antenna Using Modified Ground Plane and Unit Cell," *Microwave and Optical Technology Letters*, Vol. 56, no. 4, pp. 851–854, 2014.

[24] S. Tripathi, A. Mohan, and S. Yadav, "Ultra Wide Band Antenna Using Minkowski Like Fractal Geometry," *Microwave and Optical Technology Letters*, Vol. 56, no. 10, pp. 2273–2279, 2014.

[25] Z. Esmati, and M. Moosazadeh, "Band-Notched CPW-Fed UWB Antenna Using V-Shaped Fractal Elements," *Microwave and Optical Technology Letters*, Vol. 57, no. 11, pp. 2533–2536, 2015.

[26] S. Tripathi, A. Mohan, and S. Yadav, "Hexagonal Fractal Ultra Wide Band Antenna Using Koch Geometry with Bandwidth Enhancement," *IET Microwave and Antenna Propagation*, Vol. 8, no. 15, pp. 1445–1450, 2014.

[27] S. Tripathi, A. Mohan, and S. Yadav, "A Compact Octagonal Shaped Fractal UWB Antenna with Sierpinski Fractal Geometry," *Microwave and Optical Technology Letters*, Vol. 57, no. 3, pp. 570–574, 2015.

Fractals in Printed Antennas   33

[28] A. Falahati, M. Naghshvarian-Jahromi, and R. M. Edwards, "Bandwidth Enhancement and Decreasing Ultra-Wide Band Pulse Response Distortion of Penta-Gasket-Koch Monopole Antennas Using Compact Grounded Coplanar Waveguides," *IET Microwaves and Antenna Propagation*, Vol. 5, no. 1, pp. 48–56, 2011.

[29] H. Fallahi, and Z. Atlasbaf, "Study of a Class of UWB CPW-Fed Monopole Antenna with Fractal Elements," *IEEE Antennas and Wireless Propagation Letters*, Vol. 12, pp. 1484–1487, 2013.

[30] K. R. Chen, C. Y. D. Sim, and J. S. Row, "A Compact Monopole Antenna for Super Wide Band Application," *IEEE Antennas and Wireless Propagation Letters*, Vol. 10, pp. 488–491, 2011.

[31] J. C. Lagarias, C. L. Mallows, and A. Wilks, "Beyond the Descartes Circle Theorem," *American Mathematical Society*, Vol. 109, pp. 338–361, 2002.

[32] R. Ghatak, B. Biswas, A. Karmakar, and D. R. Poddar, "A Circular Fractal UWB Antenna Based on Descartes Circle Theorem with Band Rejection Capability," *Progress in Electromagnetics Research C*, Vol. 37, pp. 235–248, 2013.

[33] B. Biswas, R. Ghatak, A. Karmakar, and D. R. Poddar, "Dual Band Notched UWB Monopole Antenna Using Embedded Omega Slot and Fractal Shaped Ground Plane", *Progress in Electromagnetics Research C*, Vol. 53, pp. 177–186, 2014.

[34] L. Wen-Tao, H. Yong-qiang, F. Wei, and S. Xiao-Wei, "Planar Antenna for 3G/Bluetooth/WiMAX and UWB Applications with Dual Band-Notched Characteristics," *IEEE Antennas and Wireless Propagation Letters*, Vol. 11, pp. 61–64, 2012.

[35] B. S. Yildirim, B. A. Cetiner, G. Roqueta, and L. Jofre, "Integrated Bluetooth and UWB Antenna," *IEEE Antennas Wireless Propagation Letters*, Vol. 8, pp. 149–152, 2009.

[36] B. Biswas, R. Ghatak, and D. R. Poddar, "UWB Monopole Antenna with Multiple Fractal Slots for Band Notch Characteristic and Integrated Bluetooth Functionality," *Journal of Electromagnetic Waves Applications*, Vol. 29, no. 12, pp. 1593–1609, 2015.

[37] B. Biswas, R. Ghatak, and D. R. Poddar, "A Fern Fractal Leaf Inspired Wideband Antipodal Vivaldi Antenna for Microwave Imaging System", *IEEE Transactions Antennas and Propagation*, Vol. 65, no. 11, pp. 6126–6129, 2017.

Problems

2.1 What is a fractal?

2.2 Why are fractals becoming popular in antenna engineering?

2.3 What are the different types of fractals popularly used in printed antennas?

2.4 Why is bandwidth enhanced in the case of fractal-based antennas?

2.5 Who is known as the father of fractal engineering?

2.6 Can you define fractals mathematically?

Chapter 3

Journey of UWB Antennas Towards Miniaturization

3.1 INTRODUCTION

This chapter provides a thorough review on different ultra-wideband (UWB) antenna design techniques. Characteristic parameters like antenna gain, impedance matching, impedance bandwidth, and radiations pattern are discussed. Progress in different categories of antenna design for UWB systems is discussed and compared. Planar monopole UWB antennas and printed monopole UWB antennas are introduced. To mitigate interference with other existing wireless system such as WLAN, WiMAX, WiFi, and GPS satellite systems, UWB antennas with different band notching characteristic are briefly discussed. Ultra-wideband antennas also include those used for imaging applications. Therefore, we give a thorough review of different antipodal Vivaldi antenna (AVA) design techniques used for microwave imaging purposes, followed by revisiting its general performance. A chronological review of the developmental aspects of the UWB antenna is presented. Ultra-wideband technology is a system that must occupy a bandwidth of at least 500 MHz, as well as having a bandwidth of at least 20% of the center frequency [1] with a limited transmit power spectral density of −41.3 dBm/MHz. Its signals are pulse-based waveforms compressed in time instead of sinusoidal waveforms compressed in frequency. Previously, UWB radar systems were developed for military applications [2]. The year 2002 witnessed regulations regarding the frequency bands between 3.1 and 10.6 GHz by the Federal Communications Commission (FCC), United States [1]. Recently, there has been much interest in developing high data–rate ultra-wideband communication systems. Research in the area of UWB systems has generated a lot of interest among microwave engineers and scientists owing to the challenges involved in the design of the hardware components in terms of the bandwidth involved for each front-end sub-system. According to the released regulations, UWB technology, which is based on transmitting ultra-short pulses with a duration of only a few nanoseconds or less, has recently received great attention in various fields like short-range wireless communication and microwave imaging. Although UWB technology has experienced many significant developments

DOI: 10.1201/9781003389859-3

in recent years, there are still challenges in making this technology live up to its full potential. The main challenge in UWB antenna design is achieving the extremely wide impedance bandwidth while maintaining other antenna characteristics like-Omni-directional radiation pattern, constant group delay, and high radiation efficiency. This chapter reviews the methods by providing a comprehensive account of design topology and antenna performance enhancement. The chapter begins with a chronological review of the developmental phases of UWB antenna design (Section 3.2). In Section 3.3, the improvement/developmental phases of UWB planar metal plate monopoles, UWB printed monopoles, and UWB slot antennas are reviewed by using both micro-strip and CPW feeding. Band notched characteristics of single, double, and multiple notches are given because this geometrical paradigm has greatly influenced recently antenna design perspectives, illustrated in Section 3.4, and finally a brief review of high-gain antipodal Vivaldi antennas for microwave imaging systems is presented in Section 3.5.

3.2 INITIAL DEVELOPMENTAL PHASE

In 1893, Heinrich Hertz [2] used a spark discharge to produce electromagnetic waves for an experiment. It was called colored noise. The earliest antenna with wideband properties was a biconical antenna made by Lodge in 1898 [3]. Its bandwidth was mainly influenced by the ending reflection due to its limited dimensions. About 20 years after Hertz's first experiment, spark gaps and arc discharge between carbon electrodes became the dominant wave generators. Improvements are going on following the path of Carter's conical monopole antenna (1939) [4], Schelkunoff's spheroidal antenna (1941) [5], Kandoian's discone antenna (1945) [6], and Brillouin's omnidirectional and directional coaxial horn antennas (1948) [7]. All these structures are large in size and bulky in volume. Both radar and communication systems could be constructed with basic components such as pulse train generators, pulse train modulators, switching pulse train generators, detection receivers, and wideband antennas. Through the 1980s, this technology was referred to as baseband, carrier-free, or impulse. The term "ultra-wideband" was not used until 1989 [8] by the US Department of Defense. Since then, with advancements in hardware design, UWB technology has been used in many applications, such as communications, radar, automobile collision avoidance, positioning systems, and liquid level sensing. In 1990, techniques for implementing UWB signaling using low-power devices were invented. In 1994, T.E. McEwan [9] invented the micro-power impulse radar (MIR). This was the first application operating at ultra-low power (9V cell operated). It was extremely compact and inexpensive. The radar used quite sophisticated signal detection and reception methods.

In 2004, researchers felt that UWB antenna systems should efficiently transmit and receive signals using UWB technology [10]. The system transfer function was studied thoroughly to express in mathematical terms. It was found that within the UWB band, the group delay should be constant. The magnitude and phase of transfer function should be flat and linear, respectively.

Figure 3.1 and 3.2 illustrate the generalized diagram of a typical transmitting/receiving antenna in UWB radio systems. It depicts the transmitting and receiving capabilities of narrowband and broadband antenna systems. For both single-band and multiband schemes, the broadband antenna system always transmits and receives pulses much more efficiently than narrow-band antenna systems.

Polarization and orientation were both considered, along with the radiation level of UWB signals, during antenna design for efficient system realization.

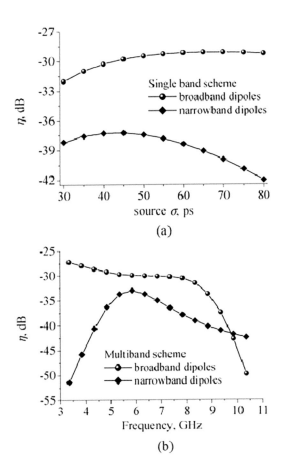

Figure 3.1 Schematic diagram of an antenna system in UWB radio system [10].
Source: © Copyright IEEE

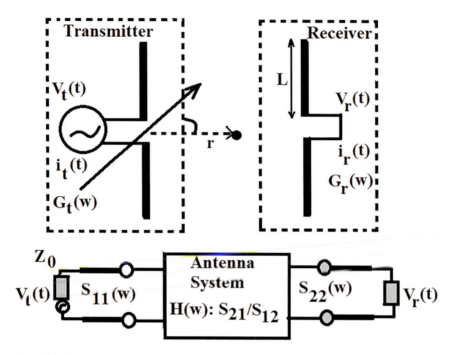

Figure 3.2 System transmission efficiency. (a) Single-band scheme. (b) Multiband scheme [10].
Source: © Copyright IEEE

3.3 IMPROVEMENT/DEVELOPMENTAL PHASE OF UWB ANTENNAS

As early as the 1970s, many new designs of planar ultra-wideband antennas had been proposed which could be classified into three categories: UWB planar metal plate monopole antennas, UWB printed monopole antennas, and UWB slot/aperture antennas.

Prior to 2005, only the conventional slot antenna was used for UWB communication systems where the antenna was omnidirectional, stable, and operated at 3.1–10.6 GHz.

3.3.1 UWB Planar Metal Plate Monopole Antennas

The main characteristic of ultra-wideband metal plate monopole antennas is that they always require a perpendicular metal ground plate. In 1976, G. Dubost [11] first invented the wideband metal plate monopole antenna, and since then, it has been continuously updated. It can be realized by replacing a conventional wire monopole with a planar monopole, where the planar monopole is located above a ground plane and commonly fed using a coaxial

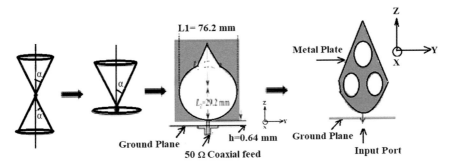

Figure 3.3 The evolution from biconical antenna to metal plate monopole antenna [12–13].
Source:© Copyright IEEE & MOTL

probe. The evolution of metal plate antennas is in chronological order from biconical antennas, discs or elliptical monopole antennas, inverted cone monopoles [12], and subsequently leaf-shaped monopole antennas [13], as shown in Figure 3.3.

The improvement of impedance bandwidth by creating different types of holes, which changes the surface current in the patch, was developed. In the early period, impedance bandwidth was only 2:1, but slowly it reached 20:1, covering the frequency range 1.3–30 GHz. Agrawal et al. [14] carried out a bandwidth comparison of several planar monopoles with various geometries, such as circular, elliptical, rectangular, and trapezoidal monopoles. The results showed that circular and elliptical monopoles exhibited much wider bandwidth performance than others. To enhance impedance bandwidth performance, several techniques, such as beveling, double feed, trident-shaped feed, and shorting, were proposed [15–18]. Ammann [15] added a bevel on one or both sides of the feed probe to increase the bandwidth with good control of the upper edge frequency in his simplest square planar monopole antenna. Another square monopole antenna with a double feed [16] exhibited enhanced impedance bandwidth and better cross-polarization levels in comparison to a square monopole antenna with a single feed. A trident-shaped feed is shown in Figure 3.4 [17]. With the use of the proposed feeding strip, the square planar monopole antenna studied shows a very wide impedance bandwidth of about 10 GHz (about 1.4–11.4 GHz). The impedance bandwidth of a wideband planar square monopole is increased dramatically by combining beveling and a shorting technique [18]. Planar antennas are designed, as they have the advantage of broad-bandwidth for impedance and radiation, stable phase response, easy fabrication and integration with other RF circuits, small size, and light weight. Due to such unique characteristics, the designs of antennas with regard to requirements vary. Subsequently, the UWB antenna is designed by optimizing for specific input signals by the use of a genetic algorithm [19].

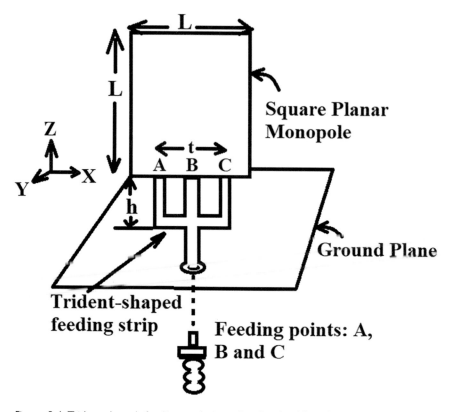

Figure 3.4 Trident-shaped feeding technique for bandwidth enhancement, Geometry of the antenna.
Source: © Copyright IEEE

The optimization technique gives the best antenna design, with high temporal correlation between electric field intensity and given input signal, with low VSWR and low dispersion over the desired frequency band. For example, in an OFDM system, the signal may be single banded and multi-banded. For a single-band impulse scheme, there are impulses with a single carrier or without a carrier that occupy the whole bandwidth. In addition, for multi-band schemes, constant gain is more important than linear phase response. To meet these criteria, antennas are modified accordingly.

3.3.2 UWB Printed Monopole Antennas

In comparison to UWB metal plate antennas, these antennas do not require a perpendicular ground plane and occupy less volume. They are suitable for integration with monolithic microwave integrated circuits (MMICs). Such antenna types consist of a monopole patch, which may be rectangular, triangular, circular, or elliptical, and a ground plane, both printed on the same plane or opposite side of the substrate. Accordingly, they are classified as micro-strip line and coplanar waveguide fed ground planes.

3.3.2.1 Microstrip Fed Printed Monopole Antennas

A circular monopole antenna was designed by Liang et al. [20] with a 50-Ω micro-strip line. It possesses an impedance ratio bandwidth of 5.3:1 with the frequency range 2.75 to 10.16 GHz, as shown in Figure 3.5(a). With an elliptical patch, the monopole antenna is also miniaturized by using an improved ground plane [21]; beveling in the ground plane is also very good for UWB performances. Among various geometrics of monopole patches, the two-step circular monopole [**Figure 3.5(b)**] is also proposed [22].

A micro-strip fed monopole antenna with rectangular patch is shown in Figure 3.6 [23–24]. The first proposed rectangular patch monopole antenna was shown by Choi et al. in 2004 [23]. It had a single slot in the patch and stair-like structure at the end of the rectangular patch, which lead to good impedance matching. A rectangular patch with two notches at the end of the two lower corners of the patch and a truncated ground plane were fabricated by Jung and Choi [24]. Lizzi et al. [25] proposed a spline-shaped monopole UWB antenna, as shown in Figure 3.6, which is capable of supporting multiple mobile wireless standards like DCS, PCS, UMTS, and ISM bands.

The triangular patch and the modified structure of the micro-strip line are shown in Figure 3.7. The antenna [26] exhibits VSWR of less than 3 for the entire frequency band of 4 to 10 GHz range. Kirchhoff's surface integral representation (KSIR) is used in the developed FDTD code to compute the

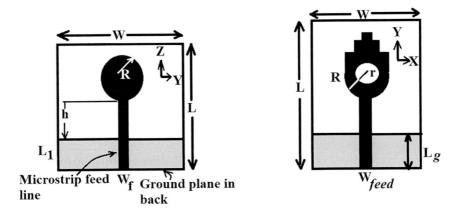

Figure 3.5 Micro-strip fed monopole antennas with circular patches [20, 22].

Source: © Copyright IEEE

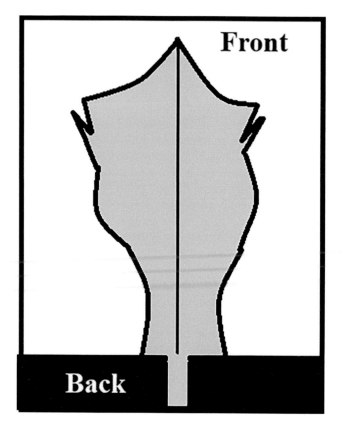

Figure 3.6 Various micro-strip fed monopole antennas with rectangular patches, Fabricated prototype.

Source: © Copyright IEEE

time-domain far-field distribution. Next, a novel triangular antenna topology based on the printed tapered monopole antenna (PTMA) is investigated in view of an ultra-wideband wireless body area network (WBAN) applications [27]. To achieve broad bandwidth, a triangular patch, truncated on the sides with modified ground plane, was designed [28].

3.3.2.2 Coplanar Waveguide Fed Printed Monopole Antennas

Fabrication of the micro-strip line is little bit complicated, as the signal line and ground planes are placed at the opposite sides of the substrate. The use of coplanar waveguide is, therefore, increasing day by day. Figure 3.8 shows different types of CPW fed printed monopole UWB antennas. Different types of ground plane, like beveling corner, circular, and trapezoidal, are used to

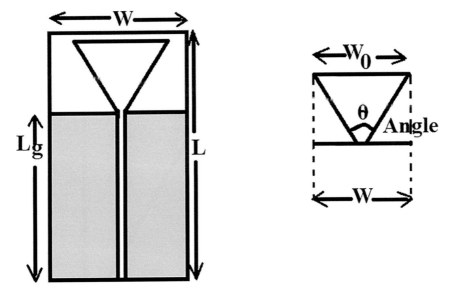

Figure 3.7 The micro-strip fed monopole antenna with triangular patch, Fabricated antenna.

Source: © Copyright IEEE

improve the matching and impedance bandwidth of the antenna [29, 30]. The antenna in Figure 3.8(b) is designed for UWB systems by using FDTD and a genetic algorithm [31].

3.3.3 UWB Slot/Aperture Monopole Antennas

Slot antennas have the advantages of low profile, light weight, wide bandwidth, and ease of fabrication. Considering these advantages, many antennas are proposed with micro-strip fed line and CPW feeding.

3.3.3.1 Microstrip Fed Slot/Aperture Monopole Antennas

Various UWB slot/aperture type antenna structures are illustrated with microstrip line feeding by a number of workers [32–39]. A tapered monopole-like slot is used instead of a rectangular slot to decrease the lower resonant frequency [32]. It also has the required bandwidth for UWB communication systems. Kalteh et al. [33] proposed a antenna with a radiator on the front side of a dielectric substrate, with a circular slot etched on this ground plane. Figure 3.9(a) shows a novel type of slot antenna [34] whose frequency characteristics can be reconfigured electronically to have either a single, dual,

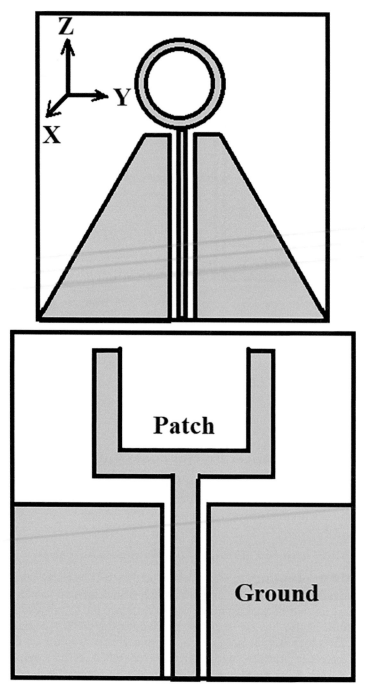

Figure 3.8 CPW fed printed monopole antennas [30, 31].
Source: © Copyright IEEE

Journey of UWB Antennas Towards Miniaturization   45

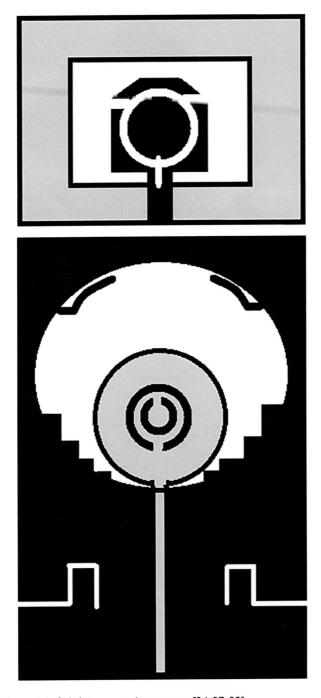

Figure 3.9 Micro-strip fed slot monopole antennas [34, 37, 39].
Source: © Copyright IEEE

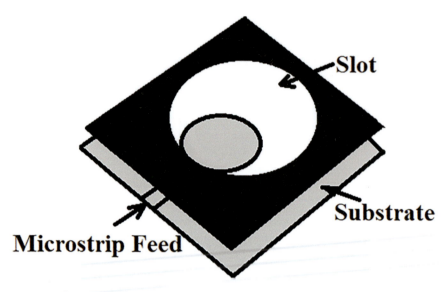

Figure 3.9 Continued

or both band notch function to block interfering signals for C-band satellite communications systems and also for IEEE802.11a and HIPERLAN/2 WLAN systems.

Different shapes of slots like hexagonal [35], printed stepped-circular [36], an semi-circle stepped aperture antennas [37], are introduced to achieve the ultra-wideband frequency range. A simple circular slot UWB antenna is designed with an impedance bandwidth beyond 175% [38], and this antenna is miniaturized by cutting in half vertically without degrading the ultra-wideband characteristics [39].

3.3.3.2 CPW Fed Slot/Aperture Monopole Antennas

Micro-strip fed slot/aperture [40, 41] antennas were discussed in the previous section. Now various types of CPW fed slot/aperture antennas [42–46] are discussed. Figure 3.10(a) presents a new design for a wide-slot antenna. The proposed printed antenna consists of a slot and a tuning stub [42]; both are formed by using a binomial curve function. Here, the order of the binomial curve function parallels the increase in bandwidth. It consists of a ground plane, shaped as semi-ellipse. Its bandwidth is 2.85–20 GHz with an omnidirectional radiation pattern [43].

A rectangular slot with a rectangular tuning stub [44] and bandwidth ratio of 1.8:1, a triangular slot with a rectangular tuning stub, and a rectangular slot with a fork-shaped structure have also been developed by Shameena et al. [45]. Gautam et al. [46] reported different types of slot antennas where the ground plane was extended toward two sides of single radiator, as shown in Figure 3.10(b).

Journey of UWB Antennas Towards Miniaturization 47

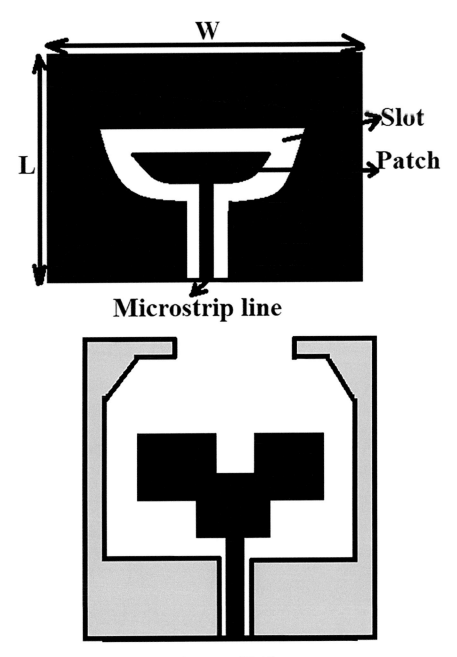

Figure 3.10 CPW-fed slot monopole antennas [42, 46].
Source: © Copyright IEEE

3.4 BAND NOTCHED CHARACTERISTICS OF UWB ANTENNAS

3.4.1 Single-Band Notched UWB Antennas

The frequency range of UWB systems interferes with other existing wireless systems such as WLAN, WiMAX, WiFi, and GPS satellite systems. To mitigate these interference issues, different types of band stop filters can be used. Nonetheless, this increases the noise level as well as the complexity of the receiver. Finally, a planar monopole antenna with band notched characteristic was designed [47]. The notch is achieved by introducing a C-shaped circle between the patch. The VSWR curve [Figure 3.11(b)] shows that it can notch out the 5.5-GHz frequency. Over time, various methods have been employed to achieve band notched characteristics such as making the slot as U shaped, arc shaped, or pie shaped on the patch followed by inserting a half- or quarter-wavelength slit on it.

A new technique has been implemented by Hong et al. [48], where a pair of T-shaped stubs were placed inside an elliptical slot, which is equivalent to a parallel LC circuit and is thus able to obtain the band-notched function. The center notch frequency and desired notch bandwidth are achieved by a properly designed equivalent capacitor and inductor values of the filter. The gain of the antenna is 4 dBi, but at notch band, it is −10 dBi. However, the efficiency decreases by 20% from the maximum efficiency of 98%.

3.4.1.1 Slots on the Radiating Patch

To obtain the frequency band-notch function in a UWB antenna, the well-known method is to insert slots on the radiating patch. Various frequency notched UWB antennas, as studied by many researchers, can be classified

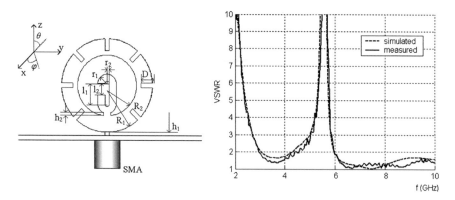

Figure 3.11 (a) C-shaped circle used for notching purposes. (b) VSWR curve shows notching at 5.5-GHz WLAN band [47].

Source: © Copyright IEEE

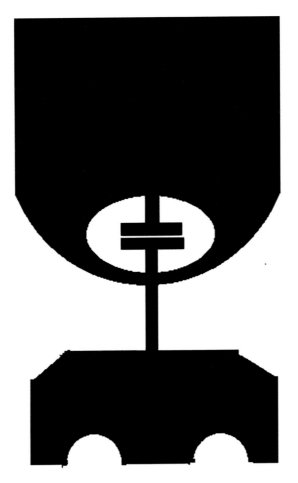

Figure 3.12 The proposed antenna.

Source: © Copyright IEEE

according to slot location, such as radiating element, ground plane, feeding line, and vicinity of the radiating element.

Different types of slots are used on the radiating patch to achieve a single notch band. In every case, the notch frequency is determined by the slot length, which is half of the guided wavelength. Other types of single-notch 5–6 GHz are achieved [49] by cutting two modified U-shaped slots with variable dimensions on the radiating patch. An additional inverted semi-elliptical patch [50] is connected to the main patch to act as a filter at the 5.1–5.9 GHz frequency. Chuang et al. [51] designed a novel second-order band stop filter composed of a non-uniform short-circuited stub and coupled open-/

short-circuited stub resonators to the fundamental antenna, as shown in Figure 3.13(c). This second-order band stop filter provides good notch-band suppressing 5.15 to 5.95 GHz, in which the normalized total radiated power in the notch-band is lower than 12 dB. Compared with the band-notched methods, using thin slits and plastic strips, this resonator has changed to a small size and a fast roll-off rate as well as 15–35 dB gain suppression.

Figure 3.13 Single frequency notched UWB antennas using slots on the radiating element [49–51].

Source: © Copyright IEEE

Journey of UWB Antennas Towards Miniaturization 51

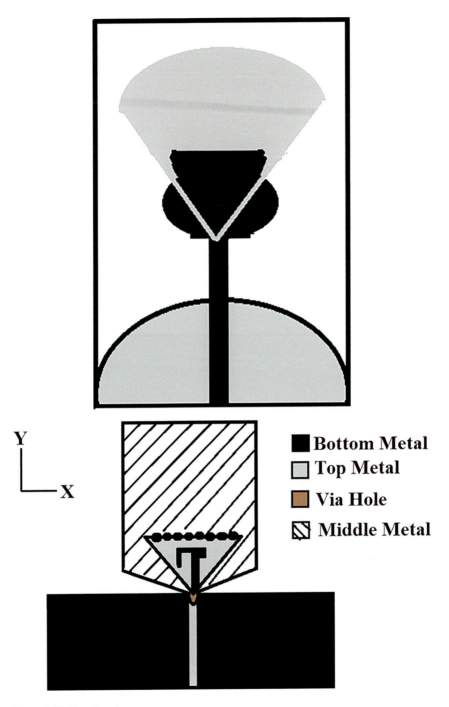

Figure 3.13 Continued

3.4.1.2 Parasitic Strips in UWB Slot Antennas

Figure 3.14 exhibits a slot antenna with single frequency notched UWB antennas using an isolated slit, open-end slits, and parasitic strips. Both of these show a CPW fed square slot UWB antenna with parasitic strip for notching a single frequency [52]. The figure also shows the utilization of two parasitic strips of half wavelength beside the patch.

3.4.1.3 Slots on the Ground Plane

Figure 3.15 exhibits different types of inserting slots in the ground plane, either in the CPW or in the micro-strip fed line, to achieve the desired single notch band. Here, in each case, the notch frequency is determined by the slot length, which is either quarter or half of the guided wavelength. In this chapter, a novel CPW fed UWB planar monopole antenna is proposed. It is accomplished by embedding a split ring resonator (SRR) array at the slot region between the antenna and ground plane of the monopole. SRRs are sub-wavelength structures that are fundamental building blocks

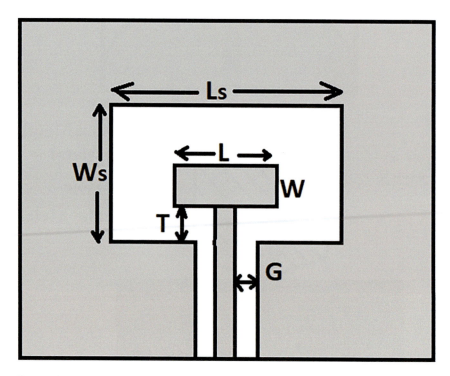

Figure 3.14 Photograph of original and three single-band notched UWB antennas.
Source: © Copyright IEEE

Journey of UWB Antennas Towards Miniaturization 53

Figure 3.15 Single-frequency notched UWB antennas using slots on the ground plane [53–54, 55].

Source: © Copyright IEEE

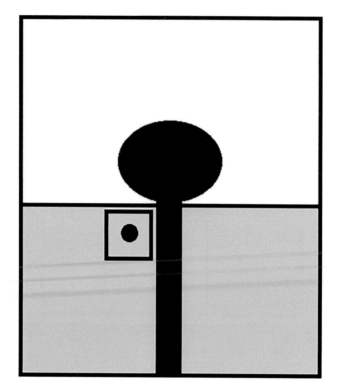

Figure 3.15 Continued

for meta-materials [53]. In Figure 3.15(b), the notch filter is achieved by a capacitively loaded loop (CLL) resonator in the vicinity of the feed line [54]. Peng and Ruan [55] utilize a mushroom-type electromagnetic band gap (EBG) structure for an effective method of band notched design.

3.4.1.4 Slots on Feed Line

Another popular position of band notch filter is in the feed line of the antenna because of high current concentration in this region. Figure 3.16(a) shows an open circuited stub connected in parallel with the micro-strip feed line [56] in order to achieve 5.5-GHz band notch performance. Figure 3.16(b) has E-shaped SIR [57] and U-shaped slots [58], respectively, in the feed line for band notching purposes.

3.4.2 Dual-Band Notched UWB Antennas

The authors are also interested in notching out two undesired frequencies simultaneously. In 2006, Lee et al. [59] designed different types of U slots, inverted U slots, L-shaped slots, and inverted L-shaped slots together

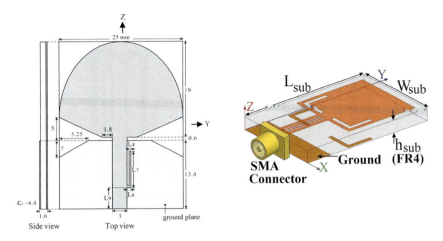

Figure 3.16 Single-frequency notched UWB antennas in the feed line [56, 57].

Source: © Copyright MOTL

with low mutual coupling between them to filter out 2.96 and 4.77 GHz simultaneously. First, a circular patch is cut out to an annular ring, and a pair of Y-shaped strips is connected to the annular ring; a notched band of WiMAX centered at 3.5 GHz is thus realized. Thereafter, an inverted V-shaped slot is etched on the patch to achieve a notched band of 5.2–5.98 GHz for the WLAN band. In Figure 3.17(b), the antenna is designed for dual band notches using two separated strips [60] for lower (5.15–5.35 GHz) and upper WLAN (5.725–5.825 GHz). By inserting a close-loop resonating structure on the radiation patch and connecting an open-loop resonator on the back side with a patch via a metallic hole, dual band [61] frequency notching is achieved. In Figure 3.17(c), two band notches were embedded in an existing UWB antenna by gluing a padding patch printed on a small single-layer piece of commercial substrate [62]. The design of the proposed new structure has controllable rejection in the 5 GHz WLAN and 8 GHz ITU frequencies. Band-notched operation is achieved by embedding novel modified CSRR slots on the radiated patch [63] [Figure 3.17(d)]. Compact short-circuited folded SIRs [64] are located beside the feed line and connected to the ground to achieve the band rejection characteristics in WiMAX and WLAN frequencies. Jiang and Che [65] designed a T-shaped stub embedded in the square slot of the radiation patch and a pair of U-shaped parasitic strips beside the feed line to achieve dual band notched characteristics.

3.4.3 Multiple-Band Notched UWB Antennas

Researchers are going on to notch out more than two frequencies simultaneously in one antenna. In Figure 3.18(a), three rectangular split ring resonators are symmetrically disposed in pairs beside the micro-strip [66] to generate

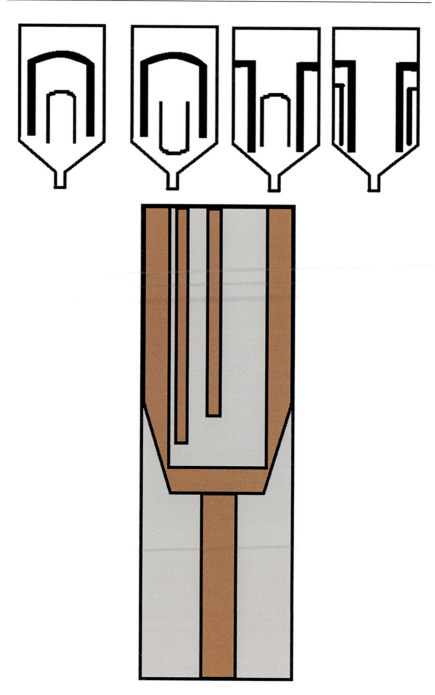

Figure 3.17 Dual-frequency notched UWB antennas [59, 60, 62–63, 65].
Source: © Copyright IEEE

Journey of UWB Antennas Towards Miniaturization 57

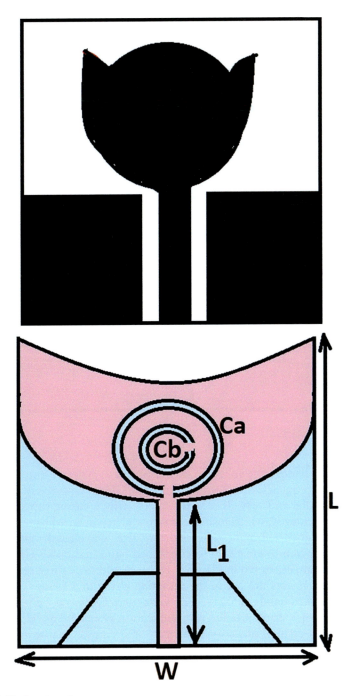

Figure 3.17 Continued

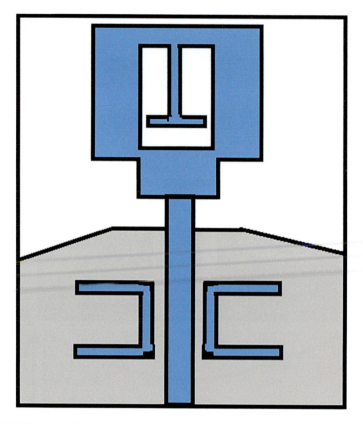

Figure 3.17 Continued

triple notched bands. The split ring slots are used to generate notched bands with central frequencies of 2.4, 3.5, and 5.8 GHz, respectively. Since the SRRs are parallel and very close to the micro-strip, a strong capacitive coupling will be achieved, which results in narrow, sharp notched bands. Nguyen et al. [67] designed a straight, open-ended quarter-wavelength slot etched in the radiating patch to create the first notched band in 3.3–3.7 GHz for the WiMAX system, as shown in Figure 3.18(b). In addition, three semicircular half-wavelength slots are cut in the radiating patch to generate the second and third notched bands in 5.15–5.825 GHz for WLAN and 7.25–7.75 GHz for downlink of X-band satellite communication systems. The band notches are realized by etching one complementary split-ring resonator inside a circular exiting stub on the front side [68]. It employs question-mark shaped (?) slot/DGS in either side of the ground plane. By adding a symmetrical pair of open-circuit stubs at the edge of the slot, tri-band rejection filtering properties in the WiMAX/WLAN are thus achieved.

A printed ultra-wideband antenna with five notched stop bands is presented by Liu et al. [69]. By introducing a pair of slots (arc-shaped), defected ground structures and an open-loop resonator, sharp frequency band-notching are observed at five different locations such as 2.4 GHz, 3.2 GHz, 5.2 GHz, 5.8 GHz, and 7.5 GHz. Figure 3.18(d) depicts it clearly.

3.5 A BRIEF REVIEW OF TAPERED SLOT ANTENNAS

The tapered slot antenna (TSA) is an extremely broadband-slot antenna wherein the slot is widened conically. The Vivaldi tapered slot antenna is a kind of aperiodic, continuously scaled, gradually curved traveling wave

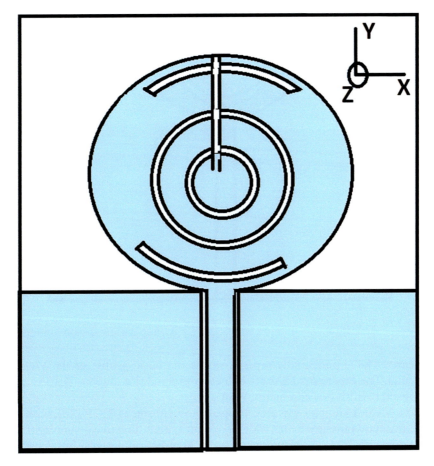

Figure 3.18 Multiple-frequency notched UWB antennas [66–68, 69].
Source: © Copyright IEEE

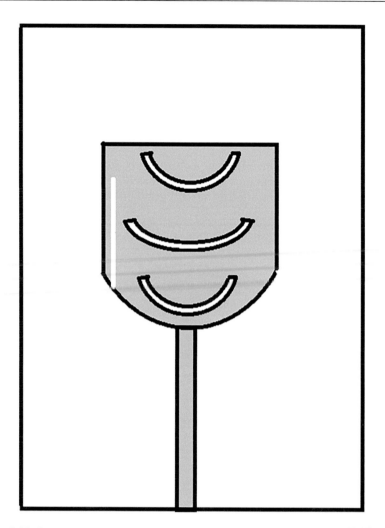

Figure 3.18 Continued

antenna. The exponentially tapered slot is the basic profile of the Vivaldi antenna. The other two TSAs are a linearly tapered slot antenna (LTSA) and a constant width slot antenna (CWSA).

The Vivaldi antenna was first proposed by Gibson in 1979 [70], with its significant advantages being wide bandwidth, high directivity, simple structure, low side lobe levels, and a symmetrical radiation pattern. In his paper, he designed a significant gain and linear polarization Vivaldi antenna with −20 dB side lobe level over an instantaneous frequency bandwidth extending from below 2 GHz to above 40 GHz.

Journey of UWB Antennas Towards Miniaturization 61

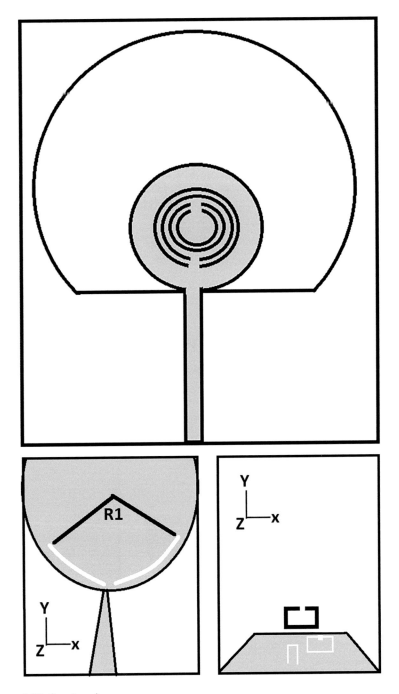

Figure 3.18 Continued

In theory, the bandwidth of a conventional Vivaldi antenna is infinite, but there are two reasons for the limitation on bandwidth of Vivaldi antennas: one is micro-strip to strip-line transition, and the other is the finite dimensions of the antenna. To solve this problem, a micro-strip to printed twin line or two-sided slot line transition was used. It generally consists of two different structures, coplanar and antipodal geometry. Compared with a coplanar Vivaldi antenna, an antipodal one gives much wider bandwidth, >10:1.

The bandwidth limitation of Vivaldi antennas is removed by the antipodal nature of the antenna, but this type of antenna provides very high levels of cross-polarization, particularly at higher frequencies due to the skew in the slot fields close to the throat of the flare. To overcome this high cross-polarization, Langley et al. [71] added a further layer of metallization to form a balanced antipodal Vivaldi antenna in 1996. The balanced antipodal transition offered a 18:1 bandwidth with fairly good cross-polarization characteristics. At different frequencies, different parts of the antipodal Vivaldi antenna radiate. In 2008, Qing et al. [72] investigated a detailed parametric study of the dual elliptically tapered antipodal slot antenna (DETSA) on impedance matching, gain and radiation patterns to provide engineers with clear designer guideline.

3.5.1 Compact Design of Vivaldi Antennas

The bandwidth of Vivaldi antennas is generally proportional to their length and aperture. Therefore, the typical DETSA becomes bulky when it is used for ultra-wideband performance. Some modifications were implemented on Vivaldi antennas to attain compact configuration [73–75], as shown in Figure 3.19.

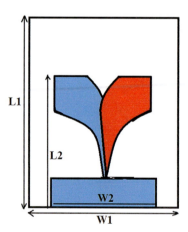

Figure 3.19 Compact antipodal Vivaldi antenna used for UWB performance [73–75].
Source: © Copyright IEEE

Journey of UWB Antennas Towards Miniaturization 63

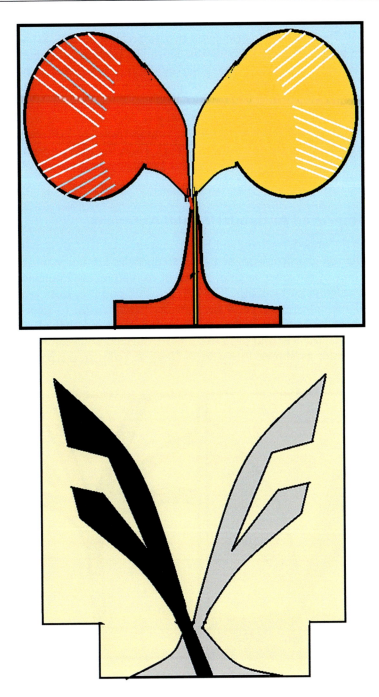

Figure 3.19 Continued

Figure 3.19(a) shows two small antipodal Vivaldi antenna [73] with dimensions of 40.16 × 42.56 mm, printed on two different substrates, RO3006 and FR4. The antenna operates across the entire UWB spectrum ranging from 3.1 to 10.6 GHz, with low cross-polarization and reasonable gain values. A novel method of improving the compactness and impedance bandwidth is proposed by Bai et al. [74]. Here two different kinds of loadings, circular load and slot load, are implemented to achieve 25:1 impedance bandwidth, as shown in Figure 3.19(b). To achieve compactness, an L-shaped slot is made at the edges of the radiating fins [75]. Thus, the lower operating frequency is increased, as shown in Figure 3.19(c), by increasing the electrical path length while maintaining the same dimensions.

3.5.2 Improved Antipodal Vivaldi Antennas

Although the impedance bandwidth of a conventional DETSA is quite wide, their radiation characteristics at higher frequencies are not stable, leading to reduction of the antenna gain. Different techniques are implemented to improve antenna performance [77–79]. For this purpose, Kota et al. [76] extended the substrate beyond the antenna metallization to improve the symmetry of its radiation patterns. This elongated and shaped substrate acts as a dielectric lens and produces a more directed radiation pattern and improved gain, as shown in Figure 3.20(a). A palm tree–type antipodal

Figure 3.20 Improved antipodal Vivaldi antenna performance [76], [77–79].
Source: © Copyright IEEE

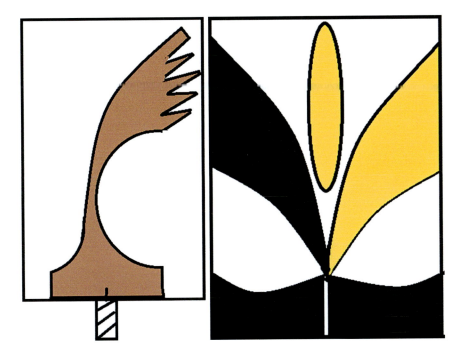

Figure 3.20 Continued

Vivaldi antenna is designed [77, 78] with improved irradiative features as compared to the conventional antipodal Vivaldi antenna, as shown in Figure 3.20(b–c). A new method for improving the directivity and bandwidth of the antipodal Vivaldi antenna structure is presented by Nassar and Weller in 2015 [79]. The authors introduce a parasitic elliptical patch in the flare aperture to enhance the field coupling between the arms and produce stronger radiation in the end fire direction [Figure 3.20(d)].

3.6 SUMMARY

In this chapter, a thorough chronological review of ultra-wideband antennas is presented. Along with an analytical introduction to UWB technology, including its advantages, properties, and so on, the preliminary developmental phase is discussed. It is seen that for general performance of UWB antennas (constant gain, impendence matching, radiation stability), the frequency notch characteristic is also required. Techniques for achieving further development and improvement of frequency notch characteristics are presented.

Different procedures have been adopted to realize UWB antennas with band-notched characteristics such as etched slots on the radiator, like U-shaped and L-shaped slots; adding a parasitic strip near the radiator; loading resonators to the feeding line; using a fractal tuning stub; and embedding a quarter-wavelength tuning stub. In earlier days, only a single-frequency notch band was designed; thereafter, dual-notched bands, triple-notched bands, and multiple-notched bands with 3.5, 5.5, 6.8, and 8.5 GHz notched frequency were designed. UWB antennas can be used in medical imaging systems, and for this highly directive and wideband antennas are required. The literature review shows different techniques used for improving the cross-polarization levels caused by antipodal Vivaldi antennas. Different modifications have been proposed for reducing antenna size and for further improvement of antenna performance, like radiation characteristics and gain.

REFERENCES

[1] A. Zanzoni, L. Montecchi-Palazzi, and M. X. Quondam, "Mint: A Molecular Interaction Database," *FEBS Letters*, Vol. 513, pp. 135–140, 2002, doi: 10.1016/s0014-5793(01)03293-8.

[2] M. Ghavami, L. B. Michael, and R. Kohno, *Ultra-Wideband Signals and Systems in Communication Engineering* (2nd edition), Hoboken: John Wiley and Sons Ltd., 2007.

[3] H. G. Schantz, "A Brief History of UWB Antennas," *IEEE Aerospace & Electronics Systems Magazine*, Vol. 19, no. 4, pp. 22–26, 2004.

[4] S. P. Carter, "Short Wave Antenna," *United States Patent*, Vol. 2, no. 175, p. 252, 1939.

[5] A. S. Schelkunoff, "Theory of Antennas of Arbitrary Size and Shape," *Proceedings of the IRE*, Vol. 29, no. 9, pp. 493–521, 1941.

[6] G. A. Kandoian, "Broadband Antenna," *United States Patent*, Vol. 2, no. 368, p. 663, 1945.

[7] N. L. Brillouin, "Broadband Antenna," *United States Patent*, Vol. 2, no. 454, p. 766, 1948.

[8] W. H. Gao, L. L. Gao, and Z. G. Liang, "A Novel Wireless Personal Area Network Technology: Ultrawide Band Technology," *Journal of Marine Science and Application*, Vol. 5, no. 3, pp. 63–69, 2006.

[9] R. Aiello, and A. Batra, *Ultrawide band Systems*, London: British Library, 2006.

[10] Z. N. Chen, X. H. Wu, H. F. Li, N. Yang, and M. Y. Wah, "Considerations for Source Pulses & Antennas in UWB Radio Systems," *IEEE Transactions on Antennas & Propagation*, Vol. 57, no. 7, pp. 1739–1748, 2004.

[11] G. Dubost, and S. Zisler, *Antennes à large bande*, New York: Masson, 1976.

[12] S. Y. Suh, W. L. Stutaman, and W. A. Davis, "A New Ultra-Wideband Printed Monopole Antenna: The Planar Inverted Cone Antenna (PICA)," *IEEE Transactions on Antennas and Propagation*, Vol. 52, no. 5, pp. 1361–1365, 2004.

[13] X. F. Bai, S. S. Zhong, and X. L. Liang, "Leaf-Shaped Monopole Antenna with Extremely Wide Bandwidth," *Microwave and Optical Technology Letters*, Vol. 48, no. 7, pp. 1247–1250, 2006.

[14] N. P. Agrawal, G. Kumar, and K. P. Ray, "Wide-Band Planar Monopole Antennas," *IEEE Transactions on Antennas and Propagation*, Vol. 46, no. 2, pp. 294–295, 1998.

[15] M. J. Ammann, "Control of the Impedance Bandwidth of Wideband Planar Monopole Antennas Using a Beveling Technique," *Microwave and Optical Technology Letters*, Vol. 30, no. 4, pp. 229–232, 2001.

[16] E. Antonino-Daviu, M. Cabedo-Fabres, M. Ferrando-Bataller, and A. Valero-Nogueira, "Wideband Double-Fed Planar Monopole Antennas," *Electronics Letters*, Vol. 39, no. 23, pp. 1635–1636, 2003.

[17] K. L. Wong, C. Wu, and S. Su, "Ultra-Wideband Square Planar Metal-Plate Monopole Antenna with a Trident-Shaped Feeding Strip," *IEEE Transactions on Antennas and Propagation*, Vol. 54, no. 4, pp. 1262–1269, 2005.

[18] M. J. Ammann, and Z. N Chen, "A Wide-Band Shorted Planar Monopole with Bevel," *IEEE Transactions on Antennas and Propagation*, Vol. 51, no. 4, pp. 901–903, 2003.

[19] T. Nikolay, and L. Yehuda, "Novel Method of UWB Antenna Optimization for Specific Input Signal Forms by Means of Genetic Algorithm," *IEEE Transactions on Antennas & Propagation*, Vol. 54, no. 8, pp. 2216–2225, 2006.

[20] J. Liang, C. C. Chiau, X. Chen, and C. G. Pirini, "Study of a Printed Circular Disc Monopole Antenna for UWB System," *IEEE Transactions on Antennas and Propagation*, Vol. 53, no. 11, 2005.

[21] C. Y. Huang, and W. C. Hsia, "Planar Elliptical Antenna for Ultra-Wideband Communications," *Electronics Letters*, Vol. 41, no. 6, pp. 296–297, 2005.

[22] A. Osama, and A. R. Sebak, "A Printed Monopole Antenna with Two Steps and a Circular Slot for UWB Applications," *IEEE Antennas and Wireless Propagation Letters*, Vol. 7, pp. 411–413, 2008.

[23] S. H. Choi, J. K. Park, S. K. Kim, and J. Y Park, "A New Ultra-Wideband Antenna for UWB Applications," *Microwave and Optical Technology Letters*, Vol. 40, no. 5, pp. 399–401, 2004.

[24] J. Jung, W. Choi, and J. Choi, "A Small Wideband Micro-Stripfed Monopole Antenna," *IEEE Microwave and Wireless Components Letters*, Vol. 15, no. 19, pp. 703–705, 2005.

[25] L. Lizzi, R. Azaro, G. Oliveri, and A. Massa, "Printed UWB Antenna Operating for Multiple Mobile Wireless Standards," *IEEE Antennas and Wireless Propagation Letters*, Vol. 10, pp. 1429–1432, 2010.

[26] C. C. Lin, Y. C. Kan, L. C. Kuo, and H. R. Chuang, "A Planar Triangular Monopole Antenna for UWB Communication," *IEEE Microwave and Wireless Components Letters*, Vol. 15, no. 10, pp. 624–626, 2005.

[27] J. R. Verbiest, and G. A. E. Vandenbosch, "A Novel Small Size Printed Tapered Monopole Antenna for UWB WBAN," *IEEE Antennas and Wireless Propagation Letters*, Vol. 5, pp. 377–379, 2006.

[28] Y. J. Cho, K. H. Kim, D. H. Choi, S. S. Lee, and S. O. Park, "A Miniature UWB Planar Monopole Antenna with 5 GHz Band Rejection Filter and the Time-Domain Characteristics," *IEEE Transactions on Antennas and Propagation*, Vol. 54, no. 5, pp. 1453–1460, 2006.

[29] W. Wang, S. S. Zhong, and S. B. Chen, "A Novel Wideband Coplanar-Fed Monopole Antenna," *Microwave and Optical Technology Letters*, Vol. 43, no. 1, pp. 50–52, 2004.

[30] X. L. Liang, S. S. Zhong, W. Wang, and F. W. Yao, "Printed Annular Monopole Antenna For Ultra-Wideband Applications," *Electronics Letters*, Vol. 42, no. 2, 2006.

[31] J. Kim, T. Yoon, J. Kim, and J. Choi, "Design of an Ultra Wide-Band Printed Monopole Antenna Using FDTD and Genetic Algorithm," *IEEE Microwave and Wireless Components Letters*, Vol. 15, no. 6, pp. 395–397, 2005.

[32] W. J. Lui, C. H. Cheng, and H. B. Zhu, "Experimental Investigation on Novel Tapered Micro-Strip Slot Antenna for Ultra-Wideband Applications," *IET Microwaves, Antennas & Propagation*, Vol. 1, no. 2, pp. 480–487, 2007.

[33] A. A. Kalteh, G. R. Dadashzadeh, M. Naser-Moghadasi, and B. S. Virdee, "Ultra-Wideband Circular Slot Antenna with Reconfigurable Notch Band Function," *IET Microwaves, Antennas & Propagation*, Vol. 6, pp. 108–112, 2012.

[34] N. Tasouji, J. Nourinia, C. Ghobadi, and F. Tofigh, "A Novel Printed UWB Slot Antenna with Reconfigurable Band Notched Characteristics," *IEEE Antennas & Wireless Propagation Letters*, Vol. 12, pp. 922–925, 2013.

[35] C. W. Zhang, Y. Z. Yin, P. A. Liu, and J. J. Xie, "Compact Dual Band Notched UWB Antenna with Hexagonal Slotted Ground Plane," *Journal of Electromagnetic Waves and Applications*, Vol. 27 no. 2, pp. 215–223, 2013.

[36] D. Abed, S. Redadda, and E. H. Kimouche, "Printed Ultra-Wideband Stepped Circular Slot Antenna with Different Tuning Stub," *Journal of Electromagnetic Waves and Applications*, Vol. 27, no. 7, pp. 846–855, 2013.

[37] X. J. Liao, H. C. Yang, N. Han, and Y. Li, "Aperture UWB Antenna with Triple Notch Characteristics," *Electronics Letters*, Vol. 47, no. 2, pp. 77–79, 2011.

[38] E. S. Angelopoulos, A. Z. Anastopoulos, D. I. Kaklamani, A. A. Alexandridis, F. Lazarakis, and K. Dangakis, "Circular and Elliptical CPW-Fed Slot and Micro-Strip-Fed Antennas for Ultra-Wideband Applications," *IEEE Antennas and Wireless Propagation Letters*, Vol. 5, no. 1, pp. 294–297, 2006.

[39] G. P. Gao, B. Hu, and J. S. Zhang, "Design of Miniaturization Printed Circular-Slot UWB Antenna by Half Cutting Method," *IEEE Antennas and Wireless Propagation Letters*, Vol. 12, pp. 567–570, 2013.

[40] P. Li, J. Liang, and X. Chen, "Study of Printed Elliptical/Circular Slot Antennas for Ultra-Wide Band Applications," *IEEE Transactions on Antennas and Propagation*, Vol. 54, no. 6, pp. 1670–1675, 2006.

[41] D. D. Krishna, M. Gopikrishna, C. K. Aanandan, P. Mohanan, and K. Vasugevan, "Compact Wide Band Koch Fractal Printed Slot Antenna," *IET Microwaves, Antennas & Propagation*, Vol. 3, pp. 782–789, 2009.

[42] X. L. Liang, T. A. Denidni, L. N. Zhang, R. H. Jin, J. P. Geng, and Q. Yu, "Printed Binomial Curve Slot Antennas for Various Wideband Applications," *IEEE Transactions on Microwave Theory and Technique*, Vol. 59, no. 4, 2011.

[43] M. Gopikrishna, D. D. Krishna, C. K. Anandan, P. Mohanan, and V. Vasudevan, "Design of a Compact Semi-Elliptic Monopole Slot Antenna for UWB System," *IEEE Transactions on Antennas and Propagation*, Vol. 57, no. 6, pp. 1834–1837, 2009.

[44] H. D. Chen, "Broadband CPW-Fed Square Slot Antennas with a Widened Tuning Stub," *IEEE Transactions on Antenna Propagation*, Vol. 51, no. 8, pp. 1982–1986, 2003.

[45] V. A. Shameena, S. Mridula, A. Pradeep, S. Jacob, A. O. Lindo, and P. Mohanan, "A Compact CPW Fed Slot Antenna for Ultra Wide Band Applications," *International Journal of Electronics and Communication* (AEÜ), Vol. 66, no. 3, pp. 189–194, 2012.

[46] A. K. Gautam, S. Yadav, and B. K. Kanaujia, "A CPW-Fed Compact UWB Micro Strip Antenna," *IEEE Antennas and Wireless Propagation Letters*, Vol. 12, pp. 151–154, 2013.

[47] J. Qui, Z. Du, J. Lu, and K. Gong, "A Planar Monopole Antenna Design with Band-Notched Characteristic," *IEEE Transactions on Antennas & Propagation*, Vol. 54, no. 1, pp. 288–292, 2006.

[48] C. Y. Hong, C. W. Ling, I. Y. Tarn, and S. J. Chung, "Design of a Planar Ultra-Wideband Antenna with a New Band-Notch Structure," *IEEE Transactions on Antennas & Propagation*, Vol. 55, no. 12, pp. 3391–3397, 2007.

[49] M. Ojaroudi, G. Ghanbari, N. Ojaroudi, and C. Ghobadi, "Small Square Monopole Antenna for UWB Applications with Variable Frequency Band Notched Function," *IEEE Antennas and Wireless Propagation Letters*, Vol. 8, pp. 1061–1064, 2009.

[50] R. Eshtiaghi, J. Nourinia, and C. Ghobadi, "Electromagnetically Coupled Band Notched Elliptical Monopole Antenna for UWB Applications," *IEEE Transactions on Antennas and Propagation*, Vol. 58, no. 4, pp. 1397–1402, 2010.

[51] C. T. Chuang, T. J. Lin, and S. J. Chung, "A Band-Notched UWB Monopole Antenna with High Notched Band Edge Selectivity," *IEEE Transactions on Antennas and Propagation*, Vol. 60, no. 10, pp. 4492–4499, 2012.

[52] Y. C. Lin, and K. J. Hung, "Compact Ultra-Wideband Rectangular Aperture Antenna and Band Notched Designs," *IEEE Transactions on Antennas and Propagation*, Vol. 54, no. 11, pp. 3075–3081, 2006.

[53] R. Ghatak, R. Debnath, D. R. Poddar, R. K. Mishra, and S. R. Bhadra Chaudhuri, "A CPW Fed Planar Monopole Band Notched UWB Antenna with Embedded Split Ring Resonators," *IEEE Antennas and Propagation Conference* (Loughborough), pp. 645–647, 2009.

[54] C. C. Lin, P. Jin, and R. W. Ziolkowski, "Single Dual and Tri Band Notched Ultrawideband (UWB) Antennas Using Capacitively Loaded Loop (CLL) Resonators," *IEEE Transactions on Antennas and Propagation*, Vol. 60, no. 1, pp. 102–109, 2012.

[55] L. Peng, and C. L. Ruan, "UWB Band Notched Monopole Antenna Design Using Electromagnetic Band Gap Structures," *IEEE Transactions on Microwave Theory and Techniques*, Vol. 59, no. 4, pp. 1074–1081, 2011.

[56] C. Y. Pan, K. Y. Chiu, J. H. Duan, and J. Y. Jan, "Band Notched Ultra Wide Band Planar Monopole Antenna Using Shunt Open Circuited Stub," *Microwave and Optical Technology Letters*, Vol. 53, no. 7, pp. 1535–1537, 2011.

[57] M. Ojaroudi, N. Ojaroudi, and S. Amiri, "Compact UWB Micro-Strip Antenna with Satellite Downlink Frequency Rejection in X-Band Communications by Etching an E-Shaped Step Impedance Resonator Slot," *Microwave and Optical Technology Letters*, Vol. 55, no. 4, pp. 922–926, 2013.

[58] L. N. Zhang, S. S. Zhong, X. L. Liang, and C. Z. Du, "Compact Omni-Directional Band-Notch Ultra-Wideband Antenna," *Electronics Letters*, Vol. 45, no. 18, pp. 659–660, 2009.

[59] W. S. Lee, D. Z. Kim, K. J. Kim, and J. W. Yu, "Wideband Planar Monopole Antennas with Dual Band Notched Characteristics," *IEEE Transactions on Microwave Theory and Techniques*, Vol. 54, no. 6, pp. 2800–2806, 2006.

[60] K. S. Ryu, and A. A. Kishk, "UWB Antenna with Single or Dual Band Notches for Lower WLAN Band and Upper WLAN Band," *IEEE Transactions on Antennas and Propagation*, Vol. 57, no. 12, pp. 3942–3950, 2009.

[61] F. C. Ren, F. S. Zhang, B. Chen, G. Zhao, and F. Zhang, "Compact UWB Antenna with Dual Band Notched Characteristics," *Progress in Electromagnetics Research Letters*, Vol. 23, pp. 181–189, 2011.

[62] M. Koohestani, N. Pires, A. K. Skrivervik, and A. A. Moreira, "Band Reject Ultra Wide Band Monopole Antenna Using Patch Loading," *Electronics Letters*, Vol. 48, no. 16, pp. 974–975, 2012.

[63] D. Jiang, Y. Xu, R. Xu, and W. Lin, "Compact Dual Band Notched UWB Planar Monopole Antenna with Modified CSRR," *Electronics Letters*, Vol. 48, no. 20, pp. 1250–1252, 2012.

[64] Y. Sung, "UWB Monopole Antenna with Two Notched Bands Based on the Folded Stepped Impedance Resonator," *IEEE Antennas & Wireless Propagation Letters*, Vol. 11, pp. 500–502, 2012.

[65] W. Jiang, and W. Che, "A Novel UWB Antenna with Dual Notched Bands for WiMAX and WLAN Applications," *IEEE Antennas & Wireless Propagation Letters*, Vol. 11, pp. 293–296, 2012.

[66] Y. Zhang, W. Hong, C. Yu, Z. Q. Kuai, Y. D. Don, and J. Y. Zhou, "Planar Ultrawide Band Antennas with Multiple Notched Bands, Based on Etched Slots on the Patch and/or Split Ring Resonators on the Feed Line," *IEEE Transactions on Antennas and Propagation*, Vol. 56, no. 9, pp. 3063–3068, 2008.

[67] T. D. Nguyen, D. H. Lee, and H. C. Park, "Design and Analysis of Compact Printed Triple Band-Notched UWB Antenna," *IEEE Antennas & Wireless Propagation Letters*, Vol. 10, pp. 403–406, 2011.

[68] X. J. Liao, H. C. Yang, N. Han, and Y. Li, "Aperture UWB Antenna with Triple Band-Notched Characteristics," *Electronics Letters*, Vol. 47, no. 2, pp. 77–79, 2011.

[69] J. J. Liu, K. P. Esselle, S. G. Hay, and S. S. Zhong, "Planar Ultra-Wideband Antenna with Five Notched Stop Bands," *Electronics Letters*, Vol. 49, no. 9, pp. 579–580, 2013.

[70] P. J. Gibson, "The Vivaldi Aerial," Proc. 9th European Microwave Conference, October, 1979, Brighton, pp. 101–105.

[71] J. D. S. Langley, P. S. Hall, and P. Newham, "Balanced Antipodal Vivaldi Antenna for Wide Bandwidth Phased Arrays," *IEE Proceedings H: Microwaves, Antennas and Propagation*, Vol. 143, no. 2, pp. 97–102, 1996.

[72] X. Qing, Z. N. Chen, and M. Y. W. Chia, "Parametric Study of Ultra Wide Band Dual Elliptically Tapered Antipodal Slot Antenna," *International Journal of Antennas and Propagation*, Vol. 2008, pp. 1–9, Article ID: 267197, 2008.

[73] A. Z. Hood, T. Karacolak, and E. Topsakal, "A Small Tapered Slot Vivaldi Antenna for Ultra Wide Band Application," *IEEE Antennas and Wireless Propagation Letters*, Vol. 7, pp. 656–660, 2008.

[74] J. Bai, S. Shi, and D. W. Prather, "Modified Compact Antipodal Vivaldi Antenna for 4–50 GHz UWB Application," *IEEE Transactions on Microwave Theory and Techniques*, Vol. 59, no. 4, pp. 1051–1057, 2011.

[75] R. Natarajan, J. V. George, M. Kanagasabai, and A. K. Shrivastav, "A Compact Antipodal Vivaldi Antenna for UWB Applications," *IEEE Antennas and Wireless Propagation Letters*, Vol. 14, pp. 1557–1560, 2015.

[76] K. Kota, and L. Safai, "Gain and Radiation Pattern Enhancement of Balanced Antipodal Vivaldi Antenna," *Electronics Letters*, vol. 47, no. 5, pp. 303–304, 2011.

[77] P. Fei, Y. C. Jiao, W. Hu, and F. S. Zhang, "A Miniaturized Antipodal Vivaldi Antenna with Improved Radiation Characteristic," *IEEE Antennas and Wireless Propagation Letters*, Vol. 10, pp. 127–130, 2011.

[78] A. M. D. Oliveira, M. B. Perotoni, S. T. Kofuji, and J. F. Justo, "A Palm Tree Antipodal Vivaldi Antenna with Exponential Slot Edge for Improved Radiation Pattern," *IEEE Antennas and Wireless Propagation Letters*, Vol. 14, pp. 1334–1337, 2015.

[79] I. T. Nassar, and T. M. Weller, "A Novel Method for Improving Antipodal Vivaldi Antenna Performance," *IEEE Transactions on Antennas and Propagation*, Vol. 63, no. 7, pp. 3321–3324, 2015.

Problems

1. What is an UWB frequency band?
2. What are the different practical usages of the UWB frequency range?
3. What are the common frequency interference issues involved in UWB range?
4. What are the different miniaturization techniques adopted in UWB antennas?
5. Why is time-domain analysis essential to characterize UWB antennas?
6. What is group-delay measurement? How is it important for UWB communication?
7. How can UWB antennas be useful for biomedical applications?
8. How the return loss graph will look alike in Smith chart?
9. What is the power spectral density in the UWB frequency band?
10. How are bandwidth widening characteristics obtained in UWB antennas?

Chapter 4

Modeling of Printed Antennas

4.1 INTRODUCTION

In the current communication world, printed antennas have taken the highest position because of their numerous merits, like light weight, small size, volume, lower packing density, economy, and above all ease of realization. They find wide application on the ground as well as in airborne devices. There are standard design formulas to construct such antennas, but, with the advent of user-friendly computational electromagnetics, optimization techniques for designing printed antennas have improved drastically. Though an efficient antenna can be designed with standard design rules and optimization techniques, it turns out to be a complicated structure. Electrical modeling can only shed some light on the way to demystify the device physics. In this chapter, three different types of printed antennas are modeled: electrically small antennas, fractal-based printed UWB antennas, and MEMS-based reconfigurable antennas. Each of these categories is further discussed for three to four varieties of printed antenna. The following three sections discuss electrical modeling. Without implementing mathematical trickery, simple RLC networks are used to model the printed antennas.

4.2 MODELING OF ELECTRICALLY SMALL ANTENNAS

While humankind is trying to achieve multi-feature communications with miniaturization, electrically small antennas (ESAs) are gaining importance compared to their counterparts. Recent years have witnessed their developmental progress because of several promising features [1–3]. Advantages of these small antennas are their radiation efficiency, physical size, and radiation quality. Though the concept of ESAs dates back to 1947–48 [4–7], their actual development started just 10 to 15 years ago because of several practical technological constraints.

Now, in the era of System-on-Chip (SoC) or Antenna-in-Package (AiP) [8, 9], ESAs are gaining popularity. This is because of its multiple favorable

DOI: 10.1201/9781003389859-4

characteristics, like light weight, compact size, and low or adequate gain for inter- or intra-chip communication. For antenna engineers, it will be comfortable to have readymade electrical models for such a small antenna structure before doing the actual 3D simulation or fabrication process.

To design ESAs efficiently, seven basic methods are generally adopted in the literature: meandering [10], looping [11], fractal implementation [12], shorting pin [13], loading of reactive elements [14, 15], slot incorporation [16, 17], and capacitive loading [18, 19]. In practice, either one method or a combination methods is used to design an ESA, but designing such antennas is a troublesome and tedious job. A circuit model approach can be implemented prior to 3D full-wave analysis or the final fabrication process steps, which can predict the trend of device performance with respect to various design parameters. Several modeling approaches have already been reported [20–22], but they deal with complicated mathematical equations. To make it easier as well as more efficient, here a simple electrical model is presented. To validate the approach, a comparison was drawn between full-wave analysis and electrical modeling for each individual case study. For proof of concept, three practical antenna problems were taken that covers a wide range of electromagnetic spectra. The first item is an on-chip antenna operating at X-band (9.45 GHz), whereas the second is a super-wideband (SWB) on-chip antenna (OCA) operating in the 2.5 to 20.6 GHz band, and the third operates in the sub-mmW range at 100 GHz.

4.3 CIRCUIT MODELING

This section outlines the electrical modeling of three different application-oriented electrically small antennas.

4.3.1 Antenna I

This section deals with the modeling of an X-band (9.45-GHz) on-chip antenna. This particular antenna is targeted for bio-telemetry applications and intra-/inter-chip communication. A standard 180-nm CMOS technology node has been used for substrate applications. It occupies an area of about 2×2.1 mm^2.

By implementing a meandering loop, capacitive loading, and shorting pin techniques, miniaturization is achieved for the structure. In terms of operating wavelength, the size of the antenna is about $\lambda_0/15$. Low-resistive silicon ($\rho = 10$ Ω-cm, $\varepsilon_r = 11.7$, tan$\delta = 0.01$) is used for the antenna structure. The geometry of the antenna is shown in Figure 4.1 with detailed dimensions, while the proposed model is depicted in Figure 4.2, with optimized parameters in Table 4.1.

Modeling of Printed Antennas 75

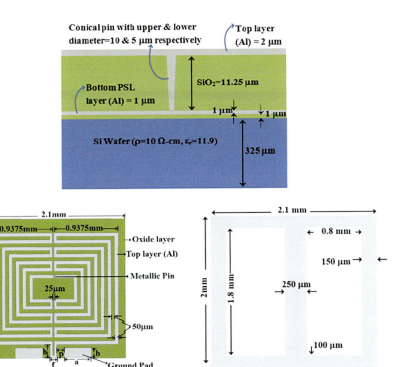

Figure 4.1 (a) Cross-sectional diagram; (b) schematic top view diagram; and (c) partially shielded layered structure at bottom of Antenna I [8].

Source: © Copyright IET-MAP

The antenna consists of three main sections: feed line (R_f and L_f), radiator part with TM_{10} resonating mode (L_{10}, C_{10}, and R_{10}), and plated through silicon oxide via-hole (C_{ox}, L_{via}, and C_{via}). Finite conductor loss and Ohmic loss are expressed by R_f and L_f, whereas the parallel RLC network or tank circuit is responsible for reserving EM energy. Preferably, the resonator is tuned at TM_{10} mode. L_{10} and C_{10} decide the resonance frequency. The parameter R_{10} dictates the bandwidth profile. Plated through via hole configuration is denoted by a parasitic lossy inductor (L_{via} and R_{via}) parallel to the oxide capacitor (C_{ox}). The C_{ox} value varies with the SiO_2 thickness.

Basically, a parallel plate capacitor is formed between the top metal layer and bottom ground plane. Via-hole inductance is a function of the radius and height of the hole. Values of all circuit elements are summarized in Table 4.1, obtained by parametric extraction techniques for this model.

Table 4.1 Optimized Parameters for the Circuit Elements of the On-Chip X-Band Antenna

Freq. (GHz)/Parameter	8	8.57	9	9.45	9.7	10	10.2	10.4	10.5	10.7
Oxide Thickness (µm)	5.25	7.25	9.25	11.25	13.25	15.25	17.25	19.25	21.25	23.25
R_{10} (Ω)	250	250	350	275	275	300	300	300	225	225
L_{10} (nH)	0.0687	0.0618	0.0618	0.054	0.0549	0.048	0.048	0.048	0.048	0.041
C_{10} (pF)	0.344	0.229	0.1486	0.344	0.147	0.44	0.39	0.24	0.19	0.73
C_{ox} (pF)	5.58	5.54	5.0678	4.99	4.922	4.97	4.82	4.77	4.7	4.7
L_{via} (pH)	0.565	0.67	0.452	0.678	0.678	0.565	0.565	0.565	0.565	0.565
R_{via} (Ω)	31.2	39	76.5	144	144.3	140	156	159	163	132
R_f (Ω)	1	3.8	0.5	0.7	1.1	1.8	1.9	2	1.3	0.5
L_f (nH)	1.085	1.085	1.085	1	1.01	1.117	1.11	1.24	1.24	1.24

Modeling of Printed Antennas 77

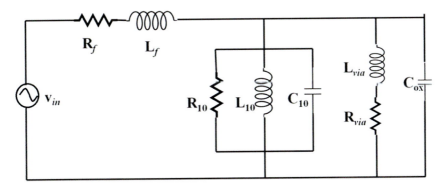

Figure 4.2 Equivalent circuit model of OCA.

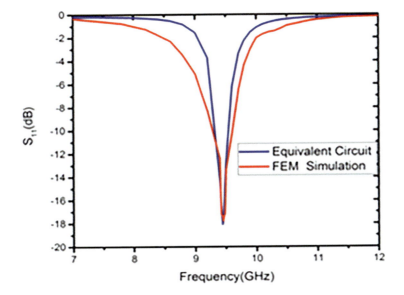

Figure 4.3 Comparative data analysis of the antenna.

Finally, Figure 4.3 shows the close relevance between the circuit modeling and full-wave analysis of the antenna structure.

4.3.2 Antenna II

This section deals with the modeling of a CPW-fed UWB type monopole antenna. Return loss bandwidth of the antenna ranges from 2.5 to 20.6 GHz, which falls under SWB range (super wide band). It has been realized on a standard 675-μm-thick high-resistive silicon ($\rho > 8$ kΩ-cm, $\varepsilon_r = 11.7$) wafer. It occupies an area of about 8.5 × 11.5 mm². A band notching feature for the 7.9- to 8.4-GHz frequency range is included in the structure to avoid the already existing ITU-T recommended X-band satellite communication (uplink). The main radiating element consists of an irregular-shaped octagonal patch with a rectangular spiral slot within it. The feed line consists of a U-shaped slot responsible for band notching behavior.

Initially, four corners are removed from the basic rectangular patch structure, and a half-wavelength ($\lambda_g/2$)–long slot is embedded with the patch. These are intended to achieve enhanced gain profile and super-wide band impedance characteristics. In the feed line, a quarter-wavelength–long U-shaped slot is included ($\lambda_g/4$) to obtain frequency notch characteristics. The geometry of the antenna is shown in Figure 4.4 with all detailed dimensions given in Table 4.2. The proposed modeling of this antenna is shown in

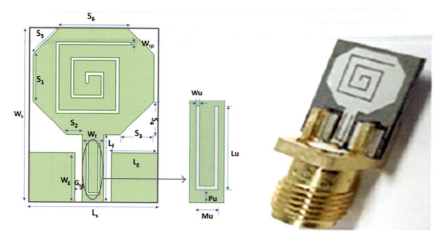

Figure 4.4 (a) Geometry of antenna II structure. (b) Prototype of Antenna-II [9].
Source: © Copyright IJMWT

Table 4.2 Optimized Parameters for Antenna-II Structure © Copyright IJMWT

Parameters	Values	Parameters	Values
L_s	8.5	P_u	0.5
W_s	11.5	W_{sp}	0.2
L_g	3.04	G_{sg}	0.51
W_g	3.3	S_1	3.1
L_f	4.5	S_2	1.1
W_f	1.4	S_3	2.2
M_u	0.9	S_4	2.2
L_u	3.4	S_5	2.4
W_u	0.15	S_6	4.6

Modeling of Printed Antennas 79

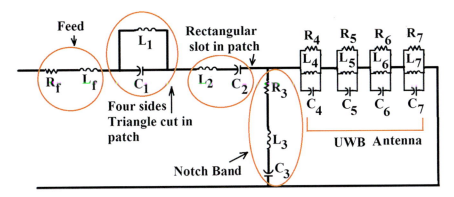

Figure 4.5 Electrical equivalent modeling of Antenna II.

Figure 4.5. Altogether it has five different segments, which explicitly describe the insight device physics. Basic UWB antenna operation can be thought as a combination of several resonant circuits, which are interlinked with each other.

It is assumed that, all of these tank circuits are operating in TM_{10} mode, whose cut-off frequency is determined by L_{10} and C_{10}, whereas, the bandwidth profile of TM_{10} mode is dictated by the R_{10} parameter. They are designated here as four different sets: (R_4, C_4, and L_4), (R_5, C_5, and L_5), (R_6, C_6, and L_6), and (R_7, C_7, and L_7).

Four corners of the patch have been cut to enhance the gain and impedance profile. These truncated corners are represented by a parallel LC network ($L_1 \| C_1$), and the spiral rectangular slot is modeled as a series combination of L and C (L_2 and C_2).

To achieve the frequency notch characteristics, a series RLC network (R_3, L_3, and C_3) is connected in parallel with the actual UWB antenna. This RLC series circuit denotes the U-shaped slot within the feed line. The corresponding inductor (L3) and capacitor (C3) values determine the notch frequency, whereas the parameter R_3 decides the notch bandwidth, Standard 50 Ω planar transmission line used for excitation of such antenna, and it is represented by series combination of R_f and L_f. Finite conductivity values are reflected by these two parameters. Table 4.3 summarizes all values of circuit elements for this model. Further, Figure 4.6 shows a close comparison of the circuit modeling and FEM simulated data.

4.3.3 Antenna III

This section outlines the modeling of a rectangular microstrip antenna array (RMAA) operating at sub-mmW range (100 GHz), specifically targeting medical imaging or THz applications. This antenna has been realized on a very thin (~100-μm) flexible substrate, called liquid crystal polymer (LCP),

Table 4.3 Equivalent Circuit Parameter Values of Antenna II

Parameters	Values	Parameters	Values
R_f	15Ω	L_4	0.75nH
L_f	1H	C_4	3.2pF
L_1	75pH	R_5	51Ω
C_1	0.65pF	L_5	0.63nH
L_2	5pH	C_5	1.08pF
C_2	100nF	R_6	63Ω
R_3	0.2Ω	L_6	0.18nH
L_3	45nH	C_6	1.028pF
C_3	8.8fF	R_7	55Ω
R_4	61Ω	L_7	0.6nH
		C_7	1.29pF

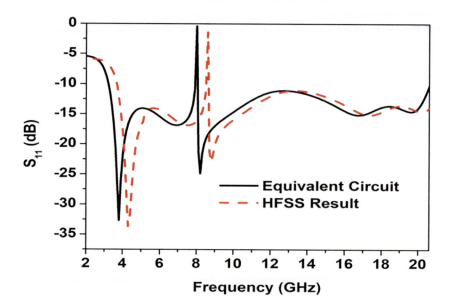

Figure 4.6 Comparison of circuit modeling and HFSS results.

which is non-toxic in nature and bio-compatible. The designed antenna occupies an area of about 12.5 × 27 mm², and because of its inherent flexible nature, it can be wrapped or conformed over cylindrical, spherical, or any regular-shaped 3D structure. Proper utilization of the 3D shape can be achieved with this antenna profile. With the implementation of parasitic patch structures along with main array elements, the peak gain of the radiator reaches 19.3 dBi, which is appreciable for such small planar

Modeling of Printed Antennas 81

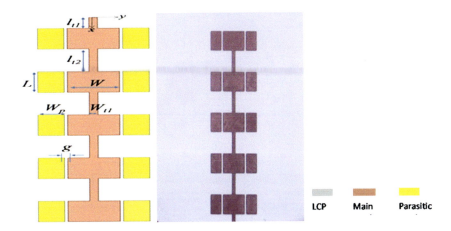

Figure 4.7 (a) Geometry of the antenna array. (b) Fabricated prototype.

Table 4.4 Optimized Dimensions of RMAA

Variables	Without Parasitic Elements (μm)	With Parasitic Elements (μm)
L	2489	2489
W	3078	3078
l_{t1}	1293	1293
l_{t2}	2685	2685
w_{t1}	690	500
w_p	–	1539
g	–	200
h	100	100

antennas. The geometry of the array with the fabricated structure is shown in Figure 4.7, with detailed dimensions as specified in Table 4.4.

This array antenna consists of five main reading elements, as shown in Figure 4.8, which can be represented by parallel tank circuits (L_{10}, C_{10}, and R_{10}). These tank circuits are assumed to resonate at TM_{10} mode. Capacitor and inductor values (C_{10} and L_{10}) decide the resonance frequency, and R_{10} determines the bandwidth profile.

Individual radiating elements are interconnected by a series-fed transmission line segment, which is expressed as a parallel combination of a lossy inductor and a capacitor. The current distribution within the line segments joining two radiating elements is expressed as L_{ij} (i = 1 to 4, j = 2 to 5)

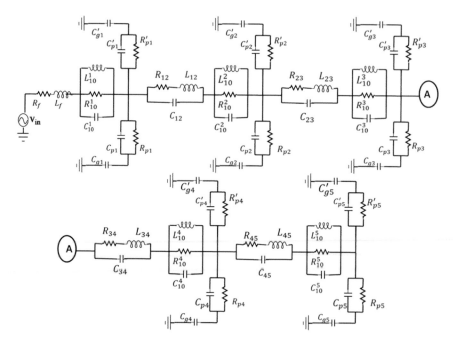

Figure 4.8 Equivalent circuit of the 100-GHz rectangular microstrip antenna array.

Table 4.5 Optimized Electrical Parameters of 100-GHz Array

Variables	Values	Variables	Values
R_f	0.01 Ω	R'_{pi}	161.12 Ω
L_f	54 pH	C_{gi}	16.45 pF
L^i_{10}	4.28 pH	C_{pi}	16.45 pF
C^i_{10}	0.591 pF	R_{pi}	161.12 Ω
R^i_{10}	44 Ω	R_{ij}	55.6 Ω
C'_{gi}	16.45 pF	L_{ij}	100 pH
C'_{pi}	16.45 pF	C_{ij}	7.4 pF

with finite conductance (R_{ij}) parasitic capacitances between two elements described by C_{ij}. Parasitic patch elements, responsible for enhancing the peak gain of the whole array, are denoted by a leaky capacitor (C_p and R_p), with an extra ground-based capacitance C_g. All the optimized parameters are summarized in Table 4.5.

A comparative data analysis of circuit modeling and full-wave FEM simulation for the 100-GHz antenna array is shown in Figure 4.9.

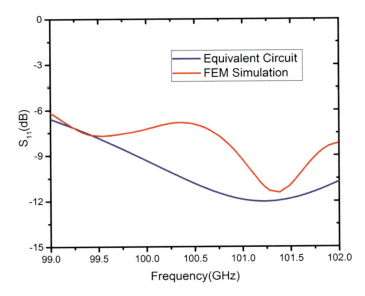

Figure 4.9 Comparative data analysis of circuit modeling and full-wave for the 100-GHz antenna array.

4.4 MODELING OF FRACTAL-BASED UWB ANTENNAS

For short-range and high–data rate indoor communication, UWB technology has gained immense popularity in the last two decades after the frequency allocation by FCC [23]. Whether civilian or strategic applications, UWB technology has spread its wings. But the challenge of future wireless communication is the manifestation of advanced features with portable or miniaturized circuits, which will consume almost negligible power. Scientists have applied their minds to learn from nature. Inspired by this, fractal engineering [24, 25] has been applied to the antenna world. It directly helps to obtain the miniaturized version of various multi-featured antennas [26–28]. 'Space-filling' and 'Self-similar' are two key features of fractal engineering, which make it an attractive choice for UWB antenna. With these, the effective length of the antenna increases, which reduces the size and enhances the impedance bandwidth by bringing the multiple resonance peaks closer. But designing such antennas is a challenging task for an RF engineer. Understanding the actual device physics is a trivial solution in this case. To alleviate these issues, a simplified electrical model of the complicated antenna structure is a great alternative for antenna engineers. Circuit modeling basically tries to explain the device physics of the antenna. With this, the importance of each design parameter can be well understood.

In this section, three fractal-based miniaturized UWB antenna problems are discussed.

4.4.1 Antenna I

This antenna uses an Apollonian fractal [29] in its circular radiator, with the added feature of a band notch at 5.5 GHz, while its operating band is a 1.8- to 10.6-GHz CPW line used for excitation. In its ground planes, two L-shaped (symmetrically placed) slots are placed to achieve band notching at 5.5 GHz. It has been realized on an FR4 substrate (ε_r = 4.4, tanδ = 0.02–0.03), and it occupies an area of about 44 × 58 mm². The geometry of the antenna is shown in Figure 4.10, with all detailed dimensions in Table 4.6. This antenna offers a peak gain of about 2 to 6 dBi (except the notch band).

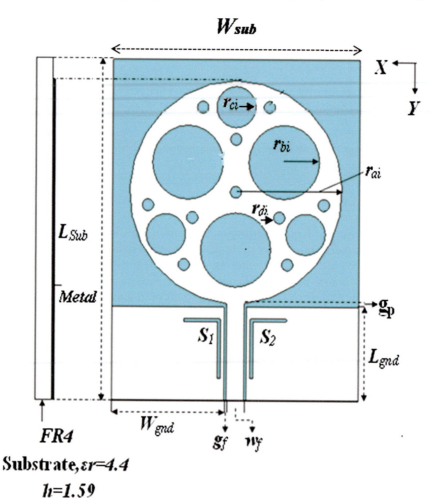

Figure 4.10 Geometry of Antenna I [29].

Source: © Copyright PIER C

Table 4.6 Detailed Dimensions of Antenna 1 [29]

Parameters	Value (mm)	Parameters	Value (mm)
L_{sub}	58	W_{gnd}	20
W_{sub}	44	$S_1(L_1 + L_2) = S_2$	16.65
g_f	0.1	R_{ai}	19
w_f	3.2	R_{bi}	3.5
L_{gnd}	15.9	R_{ci}	1.5

Source: © Copyright PIER C

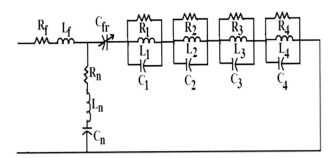

Figure 4.11 Electrical equivalent of Antenna 1.

As discussed previously, numerous advantages are obtained with the implementation of fractals in antenna geometry. Multiple resonance peaks are observed in the return loss profile, which are placed close to each other. This phenomenon can be expressed in terms of the cascaded version of tank circuits (parallel RLC networks), as shown in Figure 4.11. It is assumed that all the resonances are tuned to TM_{10} mode. A bank of parallel RLC networks (R_1 to R_4, L_1 to L_4, and C_1 to C_4) are staggered to represent these resonance phenomena. Along with this, one more tunable capacitor (C_{fr}) can be placed in series. The value of C_{fr} is changed by the fractal's iterative methods. L-shaped slots a quarter wavelength long can be modeled as a sharp band notch filter centering around 5.5 GHz. The value of this notch frequency is decided by the combination of L_n and C_n, whereas R_n determines its impedance bandwidth profile. The feed-line of the antenna is modeled as a combination of two lumped elements, L_f and R_f. The finite value of the conductance value of the metallization layer is expressed as R_f, while the change in current distribution is denoted by L_f. C_{fr} represents an Apollonian fractal for which the resonance frequency shifts down from 2 to 1.8 GHz. Optimized values of each circuit element are summarized in Table 4.7. To validate the model, comparative data has been analyzed between the full-wave analysis and performance of circuit modeling. Both data sets resemble each other well, as shown in Figure 4.12.

Table 4.7 Optimized Values of Circuit Elements of Antenna I

Parameters	Value	Parameters	Value
R_f	30 mΩ	R_2	56 Ω
L_f	100 fH	L_2	0.22 nH
R_n	0.2 Ω	C_2	7.66 pF
L_n	95 nH	R_3	55 Ω
C_n	8.8 fF	L_3	0.61 nH
C_{fr}	1 μF	C_3	1.04 pF
R_1	55 Ω	R_4	50 Ω
L_1	3 nH	L_4	0.24 nH
C_1	1.84 pF	C_4	0.95 pF

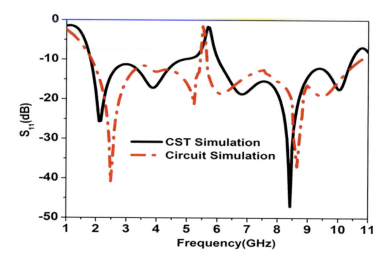

Figure 4.12 Comparison of reflection coefficient between CST and circuit simulation of Antenna I.

4.4.2 Antenna II

In this fractal-based UWB antenna, two Ω-shaped slots have been used in the main radiator to get dual band notch characteristics at 5.5 and 7.5 GHz, respectively [30]. This antenna is also excited by a standard CPW feed, whose ground planes are loaded with two crown-shaped fractal slots. The geometry of the antenna is shown in Figure 4.13, with all dimensions summarized in Table 4.8. The cross-sectional area of this structure is about 28 × 36.7 mm², and the fabricated prototype exhibits a peak gain of around 2 to 6 dBi, with a radiation efficiency of 80%, except the notch point. A proposed equivalent model of this antenna is shown in Figure 4.14.

Modeling of Printed Antennas 87

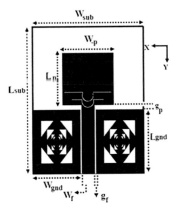

Figure 4.13 Geometry of Antenna II [30].
Source: © Copyright PIER C

Table 4.8 Detailed Dimensions of Antenna II [30]

Parameters	Value (mm)	Parameters	Value (mm)
L_{sub}	36.7	W_{gnd}	12
W_{sub}	28	W_p	13
g_f	0.4	L_p	13
w_f	3.2	g_p	1.2
L_{gnd}	16		

Source: © Copyright PIER C

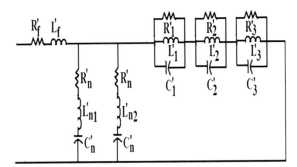

Figure 4.14 Electrical equivalent of Antenna II.

The basic UWB monopole antenna is represented by a staggered version of three parallel RLC networks. With the alteration of fractal geometry, the effective L and C values are changed for the aforementioned tank circuits, which further determines the resonant frequency peaks. It is assumed that these resonators are tuned to TM_{10} mode. The dual band notch characteristics are obtained by the inclusion of two Ω-shaped slots, which are predominantly responsible for attenuation radiation. In equivalence, it can be modeled as two series RLC networks. The notch points are achieved with the values set by the parameters L_{n1}, L_{n2}, and C_n, while the bandwidth of the notch profile is governed by the circuit's resistance value (R_b). The feed structure is symbolized as series combination of R_f and L_f. Table 4.9 summarizes the values of all circuit parameters. The validity of the proposed model is shown by comparing it with full-wave simulated results, as shown in Figure 4.15.

4.4.3 Antenna III

This is a dual-feed half annular–shaped antenna with fractal-shaped binary tree architecture. Additionally, an electromagnetically coupled disc-shaped parasitic patch is introduced at the back side of the antenna substrate to

Table 4.9 Optimized Values of Circuit Elements of Antenna II

Parameters	Value	Parameters	Value
R'_f	30 mΩ	R'_2	63 Ω
L'_f	100 fH	L'_2	0.63 nH
R'_n	0.2 Ω	C'_2	1.08 pF
L'_n	50 nH	R'_3	62 Ω
C'_n	8.8 fF	L'_3	0.21 nH
L'_{n1}	95 nH	C'_3	1.03 pF
R'_1	61 Ω		
C'_1	3.2 pF		

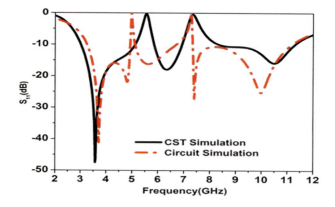

Figure 4.15 Comparison of reflection coefficient between CST and circuit simulation of Antenna II.

enhance the bandwidth of the whole antenna [31]. Three different fractal shapes have been used here to obtain the band-notching characteristics at 5.5 GHz. These are: Minkowski, Hybrid and Sierpinski fractal. In addition, a binary tree is introduced in the main antenna architecture to enhance the capability covering the Bluetooth (2.45 GHz) band. This antenna shows a peak gain of about 1 to 1.5 dBi for the Bluetooth band and 2.5 to 6 dBi for the rest of the UWB frequency range. The geometry of the antenna is shown in Figure 4.16(a and b), with detailed dimensions as in Table 4.10. The proposed model of the antenna is shown in Figure 4.17.

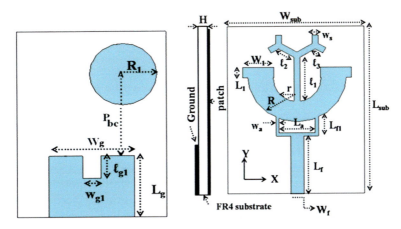

Figure 4.16 Geometry of Antenna III. (a) Front view. (b) Back view [31].
Source: © Copyright JEMWA

Table 4.10 Detailed Dimensions of Antenna III [31]

Parameters	Value (mm)	Parameters	Value (mm)
L_{sub}	33.3	ℓ_3	1.65
W_{sub}	24.4	w_s	1
w_f	2.5	W_1	6
L_f	11.35	L_1	2
L_{f1}	3.25	R_1	5.5
L_a	6.8	P_{bc}	14.65
W_a	1.02	W_g	14
R	9.5	L_{g1}	4
r	4.5	W_{g1}	3
ℓ_1	8.5	L_g	11
ℓ_2	3.3		

Source: © Copyright JEMWA

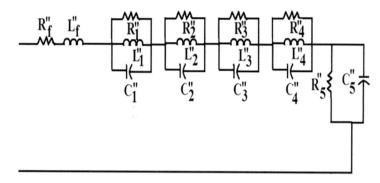

Figure 4.17 Electrical equivalent of Antenna III.

The antenna consists of four of parallel RLC networks. Multiple resonance characteristics can be explained with the tank circuits. It is assumed that all RLC networks are resonating at TM_{10} mode. The inductor and capacitor values decide the resonance frequency, whereas the resistor value determines the impedance bandwidth profile. In this antenna, the circular patch, which is placed at the back side of the substrate, can be modeled as a lossy capacitor ($C''_5 \| R''_5$). The dimensions of the patch, along with its position, decide the C''_5 values, whereas the R''_5 symbolically denotes the surface/leaky wave losses associated with the antenna structure. The feed line is represented here as a lossy inductor (R_f and L_f in series). The finite value of metal conductance is given as R_f, and the change in current distribution is expressed as L_f. Optimized values of the parameters are given in Table 4.11, with all the parameter values. The performance of the proposed model is shown with full-wave analysis of the antenna in Figure 4.18.

4.5 MODELING OF RECONFIGURABLE MEMS-BASED PATCH ANTENNAS

Modern-day communication demands portable, multi-featured, efficient circuits. With the advent of various switching technologies [32], the whole scenario is changing rapidly. The concept of a "smart antenna" was conceived only in the last decade. Re-configurability can make an antenna smart enough for practical application. This configurability can be achieved in terms of frequency, polarization, directivity, or a combination of two or three. In the last two decades, RF-MEMS technology has proved a key enabling solution in the realization of reconfigurable RF devices. Compared to existing switching technologies, RF-MEMS [33–35] technology seems superior because of its unique characteristics.

Table 4.11 Optimized Values of Circuit Elements of Antenna III

Parameters	Value	Parameters	Value
R''_f	30 mΩ	R''_3	51 Ω
L''_f	100 fH	L''_3	0.59 nH
R''_1	55 Ω	C''_3	1.0 pF
L''_1	3.18 nH	R''_4	53 Ω
C'_1	1.84 fF	L''_4	0.21 nH
R''_2	60 Ω	C''_4	1.0 pF
L''_2	0.26 nH	R''_5	13.4 Ω
C''_2	7.5 pF	C''_5	1.0 pF

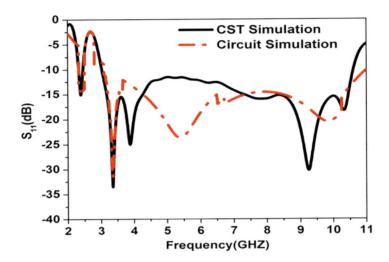

Figure 4.18 Comparison of reflection coefficient between CST and circuit simulation of Antenna III.

Micromachining techniques [36–38] can change the whole color of the antenna world in the case of system-on-chip realization. A lot of device optimization is essential prior to fabrication. Even the fabrication process also faces challenges at every step. So prior device modeling will guide a designer to choose the right path.

Here, we discuss a MEMS-based reconfigurable Ku-band antenna, whose operating frequency can alter from 14.3 to 14.5 GHz and vice-versa with the change of switching states, that is, ON/OFF. The antenna exhibits linear polarization with 4.8 dBi gain (peak). It has been realized on 675 ± 30-μm-thick high-resistive silicon ($\varepsilon_r = 11.7$, $\rho > 8$ kΩ-cm, $\tan\delta = 0.01$). Bulk and surface micromachining have both been implemented here. The following section will elaborate on the circuit modeling strategy.

4.6 CIRCUIT MODELING

Conventional microstrip antennas have numerous advantages, but they also suffer from lot of problems. MEMS can be a blessing to antenna engineers. It can alleviate various issues associated with microstrip antennas. Gain, bandwidth, radiation efficiency, or the figure-of-merit of these printed antennas can be improved significantly in the MEMS version. Printed work discusses a Ku-band reconfigurable antenna. Here, the RF-MEMS switch basically alters the electrical length of the radiating element; thus it changes the center frequency of operation. With the dynamic variation of ON/OFF [39] switching states, the center frequency changes from 14.3 to 14.5 GHz. The geometry of the antenna is shown in Figure 4.19 [40], with detailed dimensions given in Table 4.12.

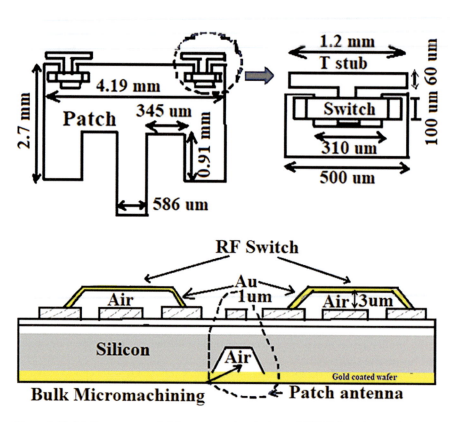

Figure 4.19 (a) Geometry and (b) cross-sectional view of MEMS-based reconfigurable antenna [40].

Source: © Copyright IRSI-13

Table 4.12 Dimensions of Proposed Antenna [40]

Parameters	Values
Length of the switch membrane	310 µm
Width of the switch membrane	100 µm
Thickness of the switch membrane	0.5 µm
Pull-down voltage	30 V
Material used for switch membrane	Gold
Initial air-gap height	3 µm

Source: © Copyright IRSI-13

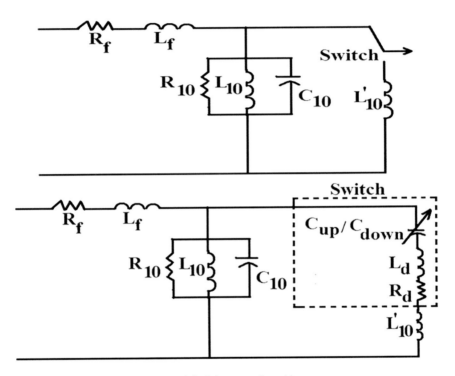

Figure 4.20 Proposed electrical model of the reconfigurable antenna.

The cross-sectional diagram of the antenna reveals the application of micromachining in the structure. A standard 675-µm-thick substrate is used as a base for the antenna. The actual radiation is positioned on a very thin silicon membrane (26 µm), which helps to improve the gain, bandwidth, and efficiency of the antenna. The proposed electrical model of the antenna is shown in Figure 4.20. This antenna can be modeled as a resonator circuit tuning in TM_{10} mode. The resonator is represented by a tank (parallel RLC network) circuit. The resonance frequency is mainly governed by the combination of L_{10} and C_{10} or ($L_{10} \parallel L'_{10}$) values. Resistance (R_{10}) determines the impedance bandwidth profile. Frequency reconfigurability is obtained by implementing a switch between L_{10} and L'_{10}. Thus, the center frequency is altered from one value to another.

From the physical geometry, it can be seen that, with the alteration of switch states, one additional conducting portion (T-shaped stub) is appended to the main patch. Thus, the electrical length changes. Input feeding is symbolized as a combination of R_f and L_f. The finite value of metal's conductivity is reflected in the non-zero value of R_f, while the change in current distribution is represented by the L_f parameter. Further, with the application of DC actuation voltage or electrostatic force, the RF-switch exhibits two different states, ON/OFF: in other words, Up/Down. The RF-MEMS switch can be modeled in terms of R, L, and C, which has been included in the antenna structure, as shown in Figure 4.21.

Here, the switch is represented by a two-state (digital) capacitor, whose value can be high or low depending upon the state of the switch. Usually at rest, it shows a lower value, and when DC voltage is applied, the capacitor value shoots up. The ratio between these two capacitors is known as the figure-of-merit (FOM), which basically decides the performance of the switch at the OPEN or DOWN state by its isolation property. Additionally, at the down state of the switch, the inductance (L_d) and contact (R_d) resistance

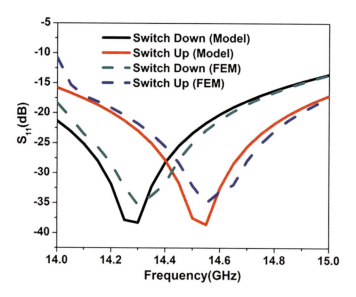

Figure 4.21 Comparison of FEM data and circuit model data.

Modeling of Printed Antennas   95

Table 4.13 Parameter Values of Equivalent Circuits of Reconfigurable Antenna

Parameter	Values	Parameter	Values
R_f	1 Ω	C_{up}	28.5 pF
L_f	1 fH	C_{Down}	2.076 fF
R_{10}	50 Ω	l_J	7.622 pH
L_{10}	0.124 nH	R_d	0.035 Ω
C_{10}	1 pF	L'_{10}	3.72 nH

parameters come into the picture. It can be clearly seen that, when the switch is closed, the overall inductance becomes a parallel combination of both L_{10} and L'_{10}, which shifts the center frequency (f_0) to 14.3 GHz. In Table 4.13, all the parameter values for circuit elements are summarized. A graph is drawn between the modeled data and full wave analysis results for two distinct states of RF switch. Close relevance is observed.

4.7 SUMMARY

This chapter shows the easiest path for beginners in this field from starting from scratch to complicated 3D model creation and optimization for printed antennas. Three popular types of printed versions of antennas were discussed here as case studies: ESA, fractal-based UWB antenna, and finally reconfigurable MEMS-based antenna.

The authors have proposed electrical models for all antenna types, which demystify the working principles of the structures in a very simple manner without applying any unwanted mathematical complexity. Comparison of the model data with the results obtained from full-wave analysis validates the proposed circuit model.

REFERENCES

[1] S. R. Best, and D. L. Hanna, "A Performances Comparison of Fundamental Small-Antenna Design," *IEEE Antennas Propagation Magazine*, Vol. 52, no. 1, pp. 47–70, 2010.

[2] O. S. Kim, O. Breinbjerg, and A. D. Yaghjian, "Electrically Small Magnetic Dipole Antennas with Quality Factors Approaching the Chu Lower Bound," *IEEE Antennas Propagation*, Vol. 58, no. 6, pp. 1898–1906, 2010.

[3] B. Biswas, R. Ghatak, and D. R. Poddar, "UWB Monopole Antenna with Multiple Fractal Slots for Band Notch Characteristic and Integrated Bluetooth Functionality," *Journal of Electromagnetic Waves and Application*, Vol. 29, no. 12, pp. 1593–1609, 2015.

[4] H. Wheeler, "Small Antennas," *IEEE Transaction of Antennas and Propagation*, Vol. 23, no. 4, pp. 462–469, 1975.

[5] H. A. Wheeler, "Fundamental Limitations of Small Antennas," *Proceeding of the IRE*, Vol. 35, no. 12, pp. 1479–1484, 1947.

[6] L. J. Chu, "Physical Limitations of Omni-Directional Antennas," *Journal of Applied Physics*, Vol. 19, pp. 1163–1175, 1948.

[7] K. Fujimoto, and H. Morishita, *Modern Small Antennas*. New York: Cambridge University Press, ISBN: 978-0-521-87786-2, 2013.

[8] H. Singh, S. Mandal, S. K. Mandal, and A. Karmakar, "Design of Miniaturized Meandered Loop On-Chip Antenna with Enhanced Gain Using Shorted Partially Shield Layer for Communication at 9.45 GHz," *IET Microwaves, Antennas and Propagation*, Vol. 13, no. 7, pp. 1009–1016, 2019.

[9] S. Mandal, A. Karmakar, H. Singh, S. K. Mandal, R. Mahapatra, and A. K. Mal, "A Miniaturized CPW-Fed On-Chip UWB Monopole Antenna with Band Notch Characteristic," *International Journal of Microwave and Wireless Technologies*, Vol. 12, no. 1, pp. 95–102, 2020.

[10] I. T. Nassar, and T. M. Weller, "An Electrically Small Meandered Line Antenna with Truncated Ground Plane," *Radio and Wireless Symposium (RWS), 2011,* doi: 10.1109/RWS.2011.5725417.

[11] M. Polívka, and A. Holub, "Electrically Small Loop Antenna Surrounded by a Shell of Concentric Split Loops," *Proceedings of the Fourth European Conference on Antennas and Propagation*, pp. 1–3, 2010.

[12] B. Biswas, R. Ghatak, and DE. R. Poddar, "A Fern Fractal Leaf Inspired Wideband Antipodal Vivaldi Antenna for Microwave Imaging System," *IEEE Transactions on Antennas and Propagation*, Vol. 65, no. 11, pp. 6126–6129, 2017.

[13] A. Chapari, A. Z. Nezhad, and Z. H. Firouzeh, "Analytical Approach for Compact Shorting Pin Circular Patch Antenna," *IET Microwaves, Antennas and Propagation*, Vol. 11, no. 11, pp. 1603–1608, 2017.

[14] R. Booket, M. Jafargholi, A. Kamyab, M. Eskandari, H. Veysi, M. Mousavi, S. Mostafa, "A Compact Multi-Band Printed Dipole Antenna Loaded with Single-Cell MTM," *IET Microwaves Antennas & Propagation*, Vol. 6, pp. 17–23, 2012.

[15] L. Wang, R. Zhang, C. L. Zhao, X. Chen, G. Fu, and X. W. Shi, "A Novel Wide Band Miniaturized Microstrip Patch Antenna by Reactive Loading," *Progress in Electromagnetic Research C*, Vol. 85, pp. 51–62, 2018.

[16] Mitra, D., Ghosh, B., Sarkhel, A. and Bhadra Chaudhuri, S. R, "A Miniaturized Ring Slot Antenna Design with Enhanced Radiation Characteristics," *IEEE Transactions on Antennas and Propagation*, Vol. 64, no. 1, pp. 300–305, 2016.

[17] R. H. Patel, A. Desai, and T. K. Upadhyaya, "An Electrically Small Antenna Using Defected Ground Structure for RFID, GPS and IEEE 802.11 a/b/g/S Applications," *Progress in Electromagnetics Research Letters*, Vol. 75, pp. 75–81, 2018.

[18] M. C. Scardelletti, G. E. Ponchak, S. Merritt, J. S. Minor, and C. A. Zorman, "Electrically Small Folded Slot Antenna Utilizing Capacitive Loaded Slot Lines," *IEEE Radio and Wireless Symposium* (Orlando), pp. 731–734, 2018.

[19] J. H. Kim, and C. H. Ahn, "Small Dual Band Slot Antenna Using Capacitor Loading," *Microwave and Optical Technology Letter*, Vol. 59, no. 9, pp. 2126–2131, 2017.

[20] C. Pfeiffer, and A. Grbic, "A Circuit Model for Electrically Small Antennas," *IEEE Transaction Antennas and Propagation*, Vol. 60, no. 4, pp. 1671–1683, 2012.

[21] A. Sohrabi, H. Dashti, and J. A. Shokouh, J. A, "Design and Analysis of a Broad Band Electrically Small Antenna Using Characteristic Mode Theory," *AEU-International Journal of Electronics and Communication*, Vol. 113, pp. 1–8, 2020, doi: 10.1016/j.aeue.2019.152991.

[22] T. L. Simpson, J. C. Logan, and J. W. Roway, "Equivalent Circuits for Electrically Small Antennas Using LS-Decomposition with the Method of Moments," *IEEE Transaction Antennas and Propagation*, Vol. 37, no. 4, pp. 462–469, 1975.

[23] First Report and Order in the Matter of Revision of Part 15 of the Commission's Rules Regarding Ultra-Wideband Transmission Systems, Released by Federal Communication Commission ET-Docket 98–153. April 22, 2002.

[24] H. Douglas, and S. Ganguly, "An Overview of Fractal Antenna Engineering Research," *IEEE Antennas and Wave Propagation Magazine*, Vol. 45, no. 1, pp. 38–57, 2003.

[25] A. Reha, O. Benhmammouch, and A. Oulad-Said, "Fractal Antennas: A Novel Miniaturization Technique for Wireless Networks," *Transactions Networks Communications*, Vol. 2, no. 5, pp. 165–193, 2014.

[26] P. Lin, R. Cheng-Li, "UWB Band-Notched Monopole Antenna Design Using Electromagnetic-Bandgap Structures," *IEEE Transactions on Microwave Theory Techniques*, Vol. 59, pp. 1074–1081, 2011.

[27] J. Liang, C. Chiau, X. Chen, and C. Parini, "Study of a Printed Circular Disc Monopole Antenna for UWB Systems," *IEEE Transactions on Antennas and Propagation*, Vol. 53, no. 11, pp. 3500–3504, 2005.

[28] B. I. Vendik, A. Rusakov, K. Kanjanasit, J. Hong, and D. Filorov, "Ultrawideband (UWB) Planar Antenna with Single, Dual and Triple Band Notched Characteristic Based on Electric Ring Resonator," *IEEE Antennas Wireless Propagation*, Vol. 16, pp. 1597–1600, 2017.

[29] R. Ghatak, B. Biswas, A. Karmakar, and D. R. Poddar, "A Circular Fractal UWB Antenna Based on Descartes Circle Theorem with Band Rejection Capability," *Progress in Electromagnetic Research C*, Vol. 37, pp. 235–248, 2013.

[30] B. Biswas, R. Ghatak, A. Karmakar, and D. R. Poddar, "Dual Band Notched UWB Monopole Antenna Using Embedded Omega Slot and Fractal Shaped Ground Plane," *Progress in Electromagnetic Research C*, Vol. 37, pp. 177–186, 2014.

[31] B. Biswas, R. Ghatak, and D. R. Poddar, "UWB Monopole Antenna with Multiple Fractal Slots for Band Notch Characteristic and Integrated Bluetooth Functionality," *Journal of Electromagnetic Waves Applications*, Vol. 29, no. 12, pp. 1593–1609, 2015.

[32] G. M. Rebeiz, *RF MEMS Theory, Design, and Technology*, Hoboken: Wiley, 2003.

[33] V. K. Varadhan, K. J. Vinoy, and K. A. Jose, *RF MEMS and Their Applications*, Hoboken: John Wiley and Sons, Ltd., 2002.

[34] J. Iannacci, "RF-MEMS for High-Performance and Widely Reconfigurable Passive Components—A Review with Focus on Future Telecommunications, Internet of Things (IoT) and 5G applications," *Journal of King Saud University-Science*, Vol. 29, no. 4, pp. 436–443, 2017.

[35] S. Lucyszyn, *Advanced RF-MEMS*, Cambridge: Cambridge University Press, 2010.

[36] G. T. A. Kovacs, N. I. Maluf, and K. E. Petersen, "Bulk Micromachining of Silicon," *Proceeding of the IEEE*, Vol. 86, no. 8, pp. 1536–1551, 1998.

[37] E. Bassous, "Fabrication of Novel Three-Dimensional Microstructures by the Anisotropic Etching of (100) and (110) Silicon," *IEEE Transactions on Electron Devices*, Vol. 25, pp. 1178–1185, 1978.

[38] J. O. Dennis, F. Ahmad, and M. H. Khir, "Advances in Micro/Nano Electromechanical Systems and Fabrication Technologies," in *Chapter-5: CMOS Compatible Bulk Micromachining*, London: InTech Open House Publication, 2013.

[39] K. Sharma, A. Karmakar, K. Prakash, A. Chauhan, S. Bansal, M. Hooda, S. Kumar, N. Gupta, and A. Singh, "Design and Characterization of RF MEMS Capacitive Shunt Switch for X, Ku, K and Ka Band Applications," *Microelectronic Engineering*, Vol. 227, 2020, Article Id: 111310, doi: https://doi.org/10.1016/j.mee.2020.111310.

[40] A. Karmakar, A. Kaur, and K. Singh, "Ku-Band Reconfigurable MEMS Antenna on Silicon Substrate," *IEEE International Radar Symposium India*, 2013 (IRSI-13), pp. 1–5.

Problems

1. What is modeling of an antenna? Why is it needed?
2. What is an electrically small antenna?
3. What are the different practical applications of small antennas?
4. What is the UWB range?
5. Write down the salient features of UWB communication.
6. Why is frequency notching required for some bands in UWB communication?
7. What is a fractal? How is it used in antenna engineering?
8. What is an on-chip antenna? Where can it be used?
9. What are inter- and intra-chip communication?
10. What is the Friis transmission formula?
11. What is a re-configurable antenna?
12. How do you make a re-configurable antenna?

Chapter 5

Printed Antennas for Biomedical Applications

5.1 INTRODUCTION

In recent years, printed antennas have found wide applications in medical science. In terms of wireless telemetry with a high or appreciable data rate, miniaturized versions of antennas are gaining popularity. Broadly, the usage of antennas in the biomedical field can be categorized as follows: diagnostic, monitoring, and therapeutic applications. Wireless transmission is most commonly performed in the 402–405-MHz band, 915-MHz range, or 2450-MHz band, which has been exclusively allocated for medical implant communication systems (MICSs), is internationally available and feasible with low power circuits, falls within a relatively low noise portion of the EM spectrum, and allows for acceptable propagation through human body. Designing antennas for biomedical usage is overall a challenging task [1–8]. A few issues have to be addressed whenever we deal with such radiators: the high permittivity of biological tissues, conductive nature of human body material, exposure or radiation limits considering human health, bio-compatibility concern, and so on; apart from these concerns, a useful link budget has to be maintained well for wireless data communication along with a high data rate of transmission.

In this chapter, the authors highlight one of the most recent topics of research on antenna engineering for biomedical applications, wireless capsule endoscopy (WCE). Currently, medical practitioners use conventional endoscopy to detect the presence of any malignant tissue within the digestive tract. Real-time images are captured during this diagnostic process, and usually it is easy to capture images of the upper part of the tract and small intestine. But this is not as much the case for the lower portion of the colon and rectum [9]. Additionally, the traditional endoscopy system is painful to patients and time consuming. To alleviate all these practical issues, wireless capsule endoscopy was convinced. This method is painless and non-invasive in nature. In this technique, a capsule is swallowed by the patient and starts to capture images of internal tissues while traverses the gastrointestinal tract (GI) under gravitational force. These real-time images are sent outside the

DOI: 10.1201/9781003389859-5

100 Printed Antennas for Wireless Communication and Healthcare

body in the form of digital data (2 images/sec). The operational life of the battery inside the capsule is 8 to 10 hours [10].

An antenna acts as an "electronic eye" for the WCE system. The dispersive nature of human tissue plays a key role in designing such antennas. In the wireless capsule endoscopy method, an important point to be considered is the omni-directional radiation pattern of the antenna, along with its broadband nature.

In this chapter, we touch upon two different schemes, elaborating upon them with a focus on WCE antenna design and development. These are:

1. Inside the capsule
2. Outside the capsule

5.2 ANTENNA DEVELOPMENT INSIDE THE CAPSULE

Though using the surface area of the capsule provides a larger gain and bandwidth for the antenna, it introduces other problems of signal interconnection in the final stage of assembly during placement of all the electronic parts inside the capsule.

5.2.1 Antenna Design

Here, the authors present a miniaturized version of the antenna which can be fitted inside a capsule in flat form instead of wrapping it inside or outside the wall of capsule in conformal form. The antenna is designed for the 915-MHz ISM band. Its size is only 7×7 mm^2, which can easily be accommodated inside the standard capsule (26×11 mm^2). The base substrate is 675-μm-thick high-resistive silicon ($\rho > 8$ KΩ-cm, $\varepsilon_r = 11.7$, tan$\delta = 0.01$); the silicon antenna paves the way for monolithic integration with other electronic components of the endoscopy capsule. Further micromachining processes can be implemented to achieve higher gain and bandwidth for this miniaturized antenna.

The proposed antenna is a slot-line fed (50-Ω) structure. The architecture of the antenna is shown in Figure 5.1.

Here, the parameter s determines the characteristic impedance of the slot line using the closed-form Equation (1) [11]. For this electrically small antenna, the overall dimension is considered one-tenth of the guided wavelength (λ_g).

$$Z_0(f, t = 0) = \frac{120\pi}{\sqrt{\varepsilon_{eff}}} \times \frac{K(k_0)}{K(k_0')} \tag{1}$$

where
$K(k)$ = complete elliptical integral of the first kind

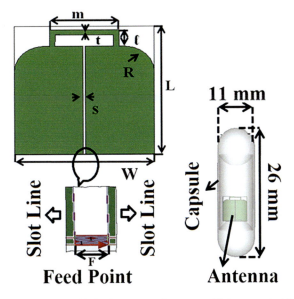

Figure 5.1 Proposed antenna. (a) Architecture of antenna. (b) Antenna inside the capsule.

K = aspect ratio of structure
ε_{eff} = Effective permittivity of structure

$$K(k') = K\left(\sqrt{1-k^2}\right)$$

As the slot-line is a balanced transmission line, to make a connection with a standard RF coaxial cable, a balanced-to-unbalanced impedance transformer, that is, a Balun circuit, is essential. The antenna is placed horizontally at one side of the inner space of the capsule, and the other side remains vacant for the placement of the camera, battery, and other components. The optimized dimensions of the antenna for 915 MHz are summarized in Table 5.1. For bio-compatibility concerns, a 100-μm-thick polyetherether ketone (PEEK; ε_r = 3.2, tanδ = 0.01) coating is chosen as the capsule shell to prevent any malignant effects with biological tissue.

For simulation purposes, the antenna is placed inside the capsule, which is ultimately positioned symmetrically within a cubical homogeneous muscle phantom (as a reference case). Similar procedures have been carried out for other human phantoms, like muscles, the stomach, and small intestines. The dimensions of the cable are chosen as 183 mm (0.5 λ_0 to 0.6 λ_0) to avoid any unwanted modes. The capsule antenna arrangement in Figure 5.1(b) was implemented at the center of the muscle tissue phantom, as shown in

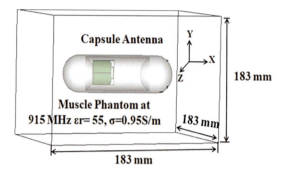

Figure 5.2 Capsule antenna inside the colon phantom.

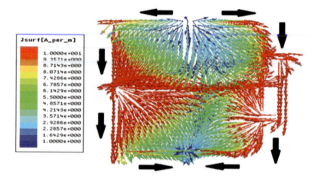

Figure 5.3 Current distribution of the proposed antenna at 915 MHz.

Figure 5.2. The surface current distribution of this electrically small antenna is shown in Figure 5.3. The maximum current inside the geometry flows in the unified direction at each end. This is the main source of radiation. At the center of the antenna, currents are out of phase with each other, so it makes hardly any contribution to antenna radiation.

5.2.2 Parametric Studies

The most important design parameter of this antenna is the slot dimension, s, and the others are chamfering radius (R), length (m), and width (t) of the upper metallic strip, as shown in Figure 5.4(b).

The slot dimension determines the 50-Ω impedance balance profile. Any major deviation from this may lead to mismatching, as shown in Figure 5.4(a), whereas Figure 5.4(b) depicts the effect of the chamfering radius on the reflection coefficient of the antenna.

Any sharp transition ($R = 0$ to a finite value) may degrade the performance metric (S_{11}) even by 50%. So an optimum value of R is chosen as 1 mm.

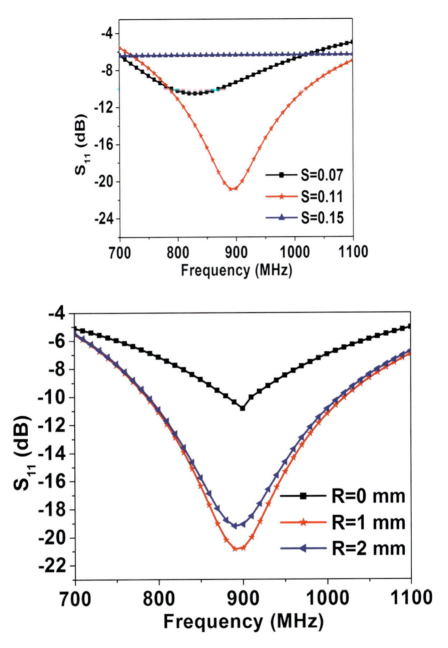

Figure 5.4 Parametric study of antenna geometry at the center for colon tissue. (a) Slot line gap dimensions. (b) Chamfering radius. (c) Length of the metal strip. (d) Width of the metal strip.

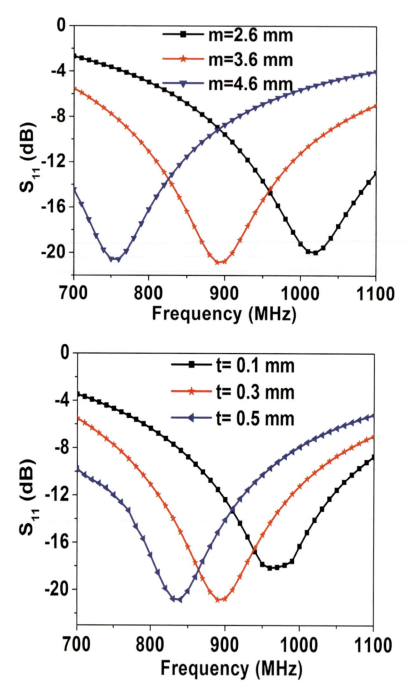

Figure 5.4 Continued

Table 5.1 Optimized Dimensions of Capsule Antenna

Parameter	Value (mm)
L	7.000
W	7.000
ℓ	0.925
m	3.600
t	0.300
s	0.110
R	1.000

The portion of the antenna (inverted U-like structure) formed by m, l, and t makes the actual radiating part. Figures 5.4(c) and 5.4(d) clearly depict the role of each design parameter. If l is constant (0.925 mm), if m is altered, then the electrical length of the antenna is changed.

As the electrical length increases, the resonant frequency decreases. For $m = 3.6$ mm, it tunes at 915 MHz. Figure 5.4(d) explains the physics of the alteration of dimension t. Due to the increase in metal strip width, the overall equivalent inductance of the antenna will increase, which leads to a lowering of the center frequency.

Initially, as a reference case, a capsule antenna is considered without any bio-compatible material or electronics components inside it. Then, it is placed at the center of the colon phantom, as shown in Figure 5.2. The simulated reflection coefficient performance of this antenna is shown in Figure 5.5. It shows an impedance bandwidth between 783 and 1015 MHz. The peak simulated gain comes out as −35.5 dBi. Further, while the capsule is coated in bio-compatible material (PEEK), it modifies the electromagnetic behavior of the radiator. Figure 5.6 shows the difference between applications of this coating material on antenna performance. Results indicate that the resonance frequency shifts slightly at the higher side due to the decrease of the dielectric frequency range, and wide bandwidth characteristics are maintained.

When a patient swallows the capsule, it traverses various biological tissues of the GI tract. Different tissues exhibit variable characteristics in the electrical domain, which can further detune the capsule antenna. So rigorous study is essential to capture this effect. Figure 5.7 shows explicitly this practical issue. The performance of the antenna changes, which is summarized in Table 5.2. This tabulated data shows the change in overall antenna performance due to changes in bio-tissue phantoms. During this downward movement of the endoscopy capsule under the influence of gravitational force, the orientation of the capsule may change randomly. But, for the case study, we have considered here only ±45° rotation with regard to the reference case's baseline in the XY, YZ, and ZX planes, as shown in Figure 5.8. Hardly any significant effect is observed here.

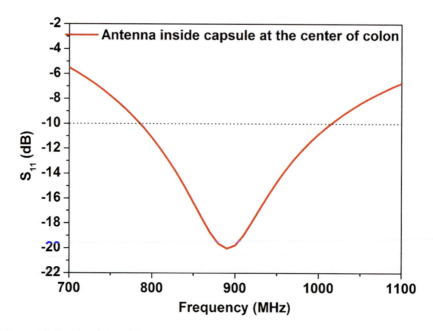

Figure 5.5 Simulated S_{11} of the capsule antenna at the center of the colon-tissue phantom.

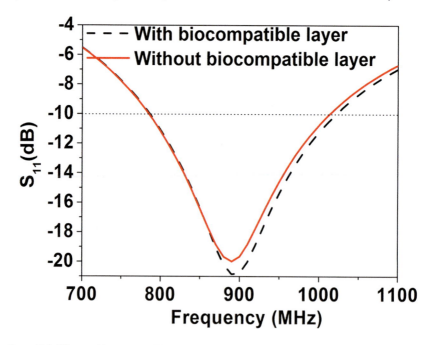

Figure 5.6 Effects of biocompatible layer on S_{11} characteristics.

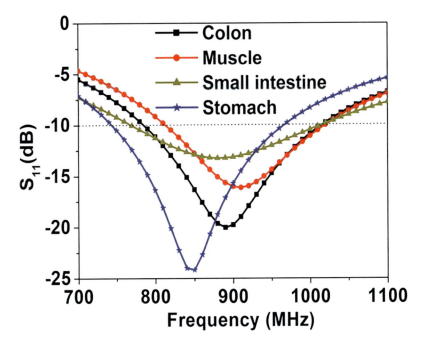

Figure 5.7 Effects on simulated S_{11} characteristics of capsule antenna with different tissue phantoms.

Table 5.2 Properties of various tissues of the human body at 915 MHz.

Tissue	Colon	Muscle	Small intestine	Stomach
ε_r	57.86	54.99	59	65
σ (S/m)	1.09	0.95	2.17	1.19
S_{11} (dB) matching	−20.02	−16.113	−13.23	−24.16
BW (MHz)	(1015 − 785) = 230	(1020 − 815) = 205	(1021 − 765) = 256	(965 − 743) = 222
Gain (dBi)	−35.5	−33	−40	−35.85
Efficiency (%)	0.03	0.06	0.01	0.03

Despite the previous points, another major point in capsule antenna design is considering the effect of the battery. As we know, an endoscopy capsule consists of different electronic components, such as camera, battery, light-emitting diodes (LEDs), and telemetry units. Among these, the battery can perturb the radiation field of the antenna significantly. The battery can be modelled as a perfect electric conductor (PEC) cylinder with 5.5 mm (max)

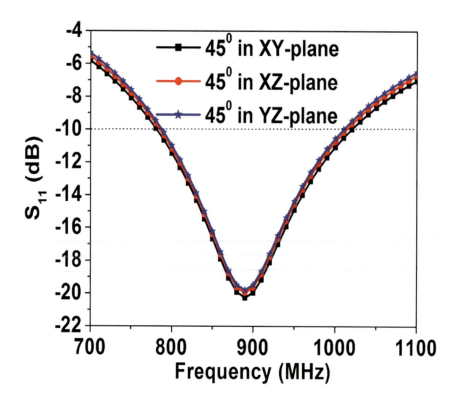

Figure 5.8 Effects on S_{11} characteristics with different capsule orientations.

diameter and 7 mm thickness. The proximity distance between the antenna and battery is also varied with the parameter D. When $D = 0$, the radiation field of the antenna is severely disturbed, which leads to severe mismatching.

At the design frequency (915 MHz), the far-field zone starts from 0.3 mm and beyond. It can be found from the standard formula of antenna calculations: $\dfrac{2D^2}{\lambda_0}$

Here, D = maximum dimension of the antenna, 0.7 mm, and λ_0 = 327 mm.

Accordingly, the battery is placed at a distance of three to four times $\dfrac{2D^2}{\lambda_0}$ Results clearly indicate this phenomenon at $D = 1$ or 2 mm, as shown in Figure 5.9.

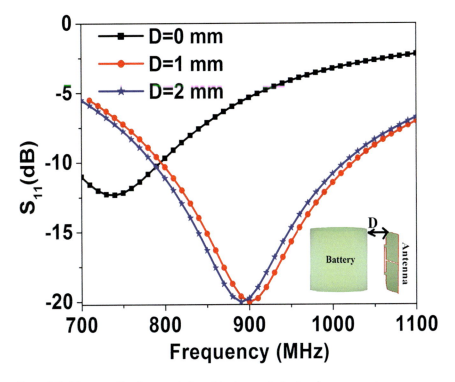

Figure 5.9 Effects on S_{11} characteristics with electronic (battery) components.

5.2.3 Measurement of the Antenna

For prototype development, the miniaturized antenna was fabricated with a 675-μm-thick HRS substrate (ε_r = 11.7, tanδ = 0.01 and ρ > 8 KΩ-cm). Initially, a stack of oxide/nitride (500Å/1500Å) was formed, followed by 1-μm aluminum sputtering. The antenna pattern was etched with wet etching chemistry. Finally, the silicon die was attached to a chip-on-board where the input connection was made through a Balun or impedance transformer. This assembly enables the user to make a comfortable connection to a standard RF co-axial cable with 50 Ω characteristic impedance. A test fixture is shown in Figure 5.10.

The simulated and measured reflection coefficient of the antenna resemble each other, as shown in Figure 5.11. Small deviations are attributed to fabrication tolerances.

An electrical model of this antenna module, along with the human phantom, is presented in Figure 5.12. Phantom liquid exhibits conductive as well

110 Printed Antennas for Wireless Communication and Healthcare

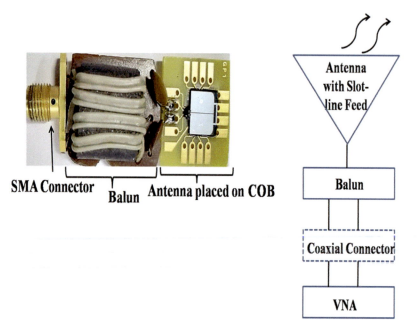

Figure 5.10 Antenna assembly.

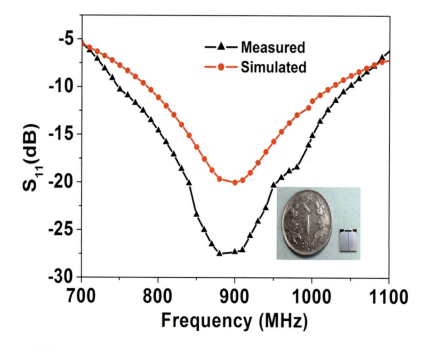

Figure 5.11 Simulated and measured reflection coefficients with fabricated prototypes.

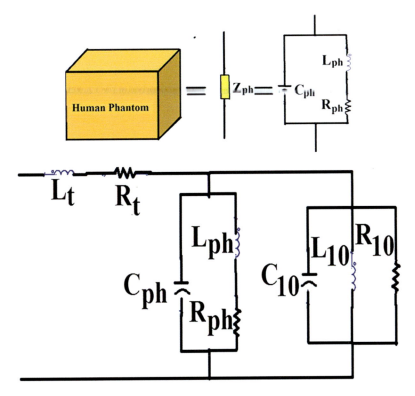

Figure 5.12 Electrical equivalent circuit. (a) Equivalent model of human phantom. (b) RLC model of antenna module.

as dielectric behavior surrounding the capsule antenna. Here, for simulation purposes, we consider a cubical shape phantom, the side of which can be represented by an equivalent impedance of Z_{ph} Ω. This impedance is composed of a parallel combination of a capacitor (C_{ph}) and lossy inductor (L_{ph} with R_{ph}). C_{ph} indicates the dielectric property of the phantom, whereas the lossy inductor denotes its conductive (electrical) nature.

The human phantom is electromagnetically coupled with the capsule antenna. The antenna is represented by a tank circuit (parallel RLC network) with its constituent parts L_{10}, C_{10}, and R_{10}. It is assumed that the antenna is tuned at TM_{10} mode. There is a flux linkage between L_{10} and L_{ph}, which reflects the effect of the human phantom on the actual antenna's performance alteration. L_{10} and C_{10} together decide the resonance frequency, while the impedance bandwidth profile is governed by R_{10}. The feed portion is represented by R_t and L_t. The finite conductivity values of the metal strip give a non-zero magnitude of R_t, and the variation of current crowding is symbolically expressed in terms of L_t.

112 Printed Antennas for Wireless Communication and Healthcare

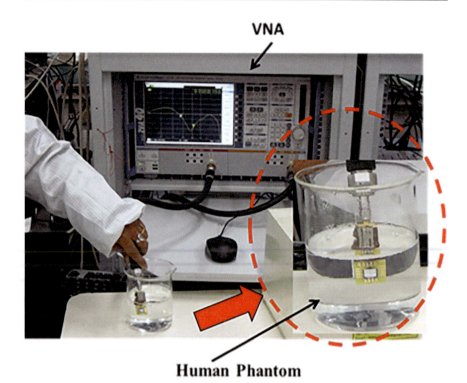

Figure 5.13 Measurement setup for fabricated antenna.

To characterize the antenna, in vitro measurement was performed with the help of a vector network analyzer (VNA), R&S-ZVA-40, and a cylindrical glass beaker full of phantom liquid, as shown in Figure 5.13.

The beaker size was chosen so that it matched well with the simulated phantom size/dimensions. For initial testing purposes, a phantom colon was made by mixing 24% (by volume) TritonX-100 with 76% salted aqueous solution (Nacl+$H_2$0) at 40°C. The salted solution was made with solid contents of 5 gm/L. TritonX-100 is a surfactant which is non-toxic in nature, with a density of 1.07 gm/cc, molar mass of 647 gm/Mol, r.i. of 1.49, and melting point of 279K. A local hot plate arrangement was created near the test setup. This was to maintain the in-situ temperature of the experimental setup.

The properties of the human phantom play a crucial role, which needs a thorough theoretical understanding and practical characterization.

A quad Cole-Cole model, shown in Equation (2), provides a parametric representation of the basic four types of dispersions in biological tissues, α, β, γ, and δ.

$$\varepsilon_r(f) = \varepsilon_r^{'}(f) - j\varepsilon_r^{''}(f) \qquad (2)$$

where

$$\varepsilon_r'(f) = \varepsilon_{s,1} + \Sigma_{l=1}^4 \frac{\varepsilon_{s,l+1} - \varepsilon_{s,l}}{1 + \left(\dfrac{jf}{f_{r,l}}\right)^{1-\alpha_l}}$$

$$\varepsilon_r''(f) = \frac{\sigma_{DC}}{2\pi f \varepsilon_0}$$

ε_r, ε_r', and ε_r'' are the complex relative permittivity, relative permittivity, and loss factor, respectively. is the DC conductivity, and and are the lth static permittivity, relaxation frequency, and distribution parameter in each dielectric dispersion, respectively.

For colon tissues, the values of $\varepsilon_r'(f)$ and $\varepsilon_r''(f)$ can be determined as follows:

$$\varepsilon_r'(f) = 4.56 + \frac{51.5}{1 + \left(\dfrac{jf}{17.6 \times 10^9}\right)^{0.964}} + \frac{2720}{1 + \left(\dfrac{jf}{18 \times 10^6}\right)^{0.842}} \tag{3}$$

$$\varepsilon_r''(f) = \frac{0.309}{2\pi f \varepsilon_0}$$

$\varepsilon_0 = 8.85 \times 10^{-12}$ F/m

The dielectric property of the human phantom was measured using Keysight's E4991 A impedance/material analyzer instrument using a dielectric probe kit. Measured and simulated data were compared, as shown in Figure 5.14. No significant difference was observed. Only 0.3% (max) deviation was noticed. This difference is attributed to human errors associated with the preparation of the homogeneous liquid phantom as compared to ideal colon tissue (as used in FEM simulation).

To validate the design with a much more practical scenario, in the FEM environment, an actual human body model was used. Likewise, the Male Torso model available in ANSYS HFSS software with volume $290 \times 230 \times 100$ mm^3 was used for simulation, as shown in Figure 5.15. It summarizes two different environments for the antenna: homogeneous and inhomogeneous human phantoms.

The E-plane pattern is like a figure eight, whereas the H-plane is almost omnidirectional. This is essential for efficient data transfer during various positions of the capsule, along with its random orientation. The cross-polarization level is around 40 dB less than the co-pole level. In an actual human body model, the smoothness of the contour of the radiation pattern is

114 Printed Antennas for Wireless Communication and Healthcare

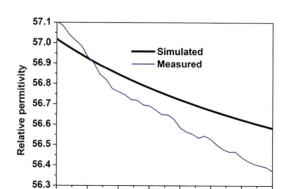

Figure 5.14 Simulated and measured phantom properties.

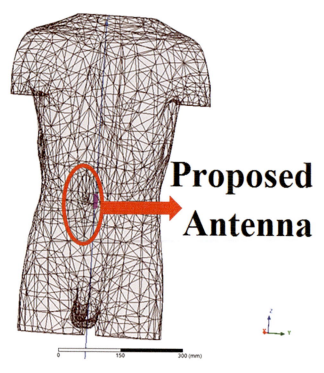

Figure 5.15 Realistic human body model. (a) 3D realistic human body model. (b) Simulated S_{11} of both single-layer colon phantom and Torso body model.

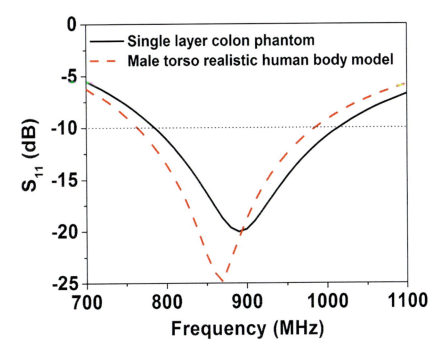

Figure 5.15 Continued

slightly disturbed due to asymmetrical changes in the dielectric and conductive properties in this case. The corresponding peak gain values are −35.5, −33, −40, and −35.85 dBi, respectively, for colon, muscle, small intestine, and stomach. This indicates that the peak gain value is a strong function of the phantom property. Further, budget analysis can be performed with this gain specification and other relevant parameters using the Friis transmission formula. The effect of battery cells can be seen clearly in the antenna's radiation pattern.

If there is a source of EM energy in the vicinity of the human body, then we should always consider the specific absorption rate (SAR) to judge the amount of energy absorbed by the unit kilogram of bio-tissue. Generally, worldwide, there are two standards per the IEEE C95.1–2005 standard, the 1 g average and 10 gm average. Limiting values of these standards are 1.6 and 2 W/Kg, respectively. To meet the regulation, the allowable power is 3.2 mW for 1 g average SAR (Table 5.3). This value is much higher than the required transmitted power (25 μw). It means the SAR should not be a problem for this capsule antenna.

116 Printed Antennas for Wireless Communication and Healthcare

Table 5.3 Maximum Power Limit for Satisfying the Standard SAR in the Human Body at 915 MHz

Antenna Locations	1-g SAR power limit (mW)
Muscle	3.2
Stomach	9.5
Colon	8
Small Intestine	6.3

5.3 ANTENNA DEVELOPMENT OUTSIDE THE CAPSULE

Mainly to save inner space for other electronic components within the capsule geometry, conformal antennas are the preferred choice to wrap outside the capsule wall.

In this section, a wideband thin (100-µm) conformal fractal-inspired miniaturized antenna is presented. The targeted frequency of operation is 2450 MHz (ISM band) [12–17]. Second, an iterated Minkowski fractal architecture is implemented here to obtain a wideband profile with a miniaturized version of the antenna. The outer wall is a prime choice to get the maximum directive gain and radiation efficiency.

5.3.1 Antenna Design

The whole design is accomplished by considering the standard dimensions of an endoscopy capsule, that is, 26×11 mm². It is a CPW-fed antenna structure, designed and developed on a 4-mil-thick flexible substrate LCP with a dielectric constant of 2.9 and loss tangent of 0.002. It is a non-toxic substrate and bio-compatible in nature. The geometry of the antenna is shown in Figure 5.16. Its optimized dimensions are shown in Table 5.4. The G/W/G dimensions of the CPW transmission line determine its characteristic impedance per the standard formula [11], as follows.

$$Z_{ocp} = \frac{30\pi}{\sqrt{\varepsilon_e}} \times \frac{K'(k)}{K(k)}$$

(4)

where

$$k = \frac{a}{b} \sqrt{\frac{1 - \left(\dfrac{b^2}{c^2}\right)}{1 - \left(\dfrac{a^2}{c^2}\right)}}, \quad K'(k) = k\left(\sqrt{1 - k^2}\right)$$

Printed Antennas for Biomedical Applications 117

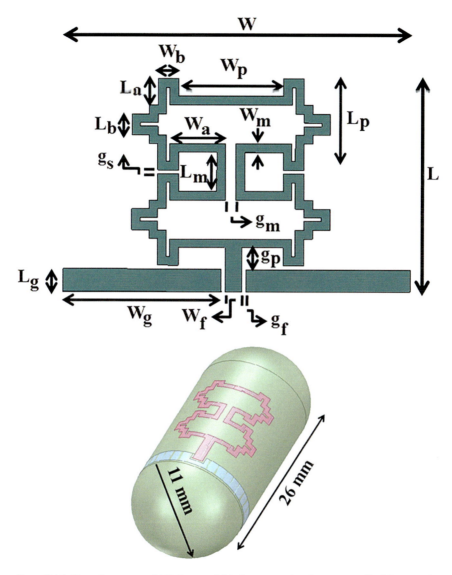

Figure 5.16 Capsule antenna. (a) Before and (b) after wrapping the outer wall of the capsule.

$a = W_f/2$; $b = a + g_f$; $c = g_f + 2.5\ W_f$
ε_e = effective permittivity of the substrate
$K(k)$ = complete elliptical integral of the first kind
k = aspect ratio

Here, fractal geometry is adopted for the miniaturization procedure, along with the wideband characteristics of the antenna. Implementation of

Table 5.4 Optimized Dimensions of the Capsule Antenna

Parameter	Value (mm)
W	20
L	13
W_g	9
L_g	1.3
W_f	1
g_f	0.2
g_p	1.2
g_m	0.6
g_s	0.2
W_b	1.2
W_p	6
L_a	1.5
L_b	1.2
L_m	2.2
W_a	2.2
W_m	0.5
L_p	5.2

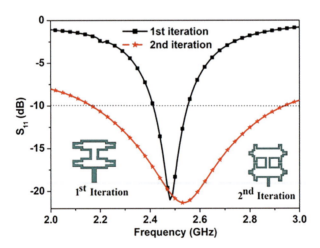

Figure 5.17 Simulated reflection coefficient of the capsule antenna at different iterations in flat form.

a second iterated Minkowski fractal in this antenna yields an impedance bandwidth which is 24.5% higher than that of the first iteration, as shown in Figure 5.17. Generally, the effective electrical length of the antenna is increased with the number of iterations applied in antenna circuits. Theoretically, though, a much higher number of iterations is possible. But, practically,

Printed Antennas for Biomedical Applications   119

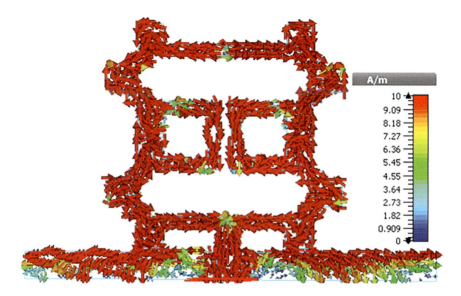

Figure 5.18 Current distribution of the conformal antenna at 2.45 GHz.

it has some limitations from the fabrication point of view. The critical dimensions of the antenna layout become infeasible in the case of larger iterations for realization in standard PCB etching chemistry. As the effective length is enhanced, so the lower cut-off frequency shifts downward (or much lower), and the upper cut-off frequency shifts towards higher values. Hence, the dynamic range or operational bandwidth is increased significantly.

Careful design also considers the surface current distribution of the antenna (Figure 5.18). It indicates smooth flow of current vectors towards the periphery of antenna, which are responsible for stable and continuous radiation.

As discussed earlier, the human body or biological tissue plays a very important role in designing any ingestible or implantable antenna. This is mainly because of the dispersive nature of the tissue. As the current antenna resides at the outer wall of the capsule, it needs more emphasis on the resonance behavior of tissue layers. So, here, in full-wave analysis, an eigenmode simulation was carried out. We consider a human phantom of cubical size $121 \times 121 \times 121$ mm^3 to predict the frequency characteristics of the phantom. Table 5.5 summarizes the eigenmode simulation results. It indicates that the resonance frequency ranges from 241 to 314 MHz, which is far from the antenna's operating frequency (2450 MHz). So there is hardly any chance of interference between these two modes. Figure 5.19 depicts the spatial variation of the E-field inside the phantom volume.

Further, the capsule antenna is placed inside a human phantom (ref. case-muscle phantom) of 121 mm^3 dimensions. This is equivalent to one free

Table 5.5 Eigenmode Simulation for Human Phantom

Tissue	Resonant Frequency (MHz)
Skin	241.168
Muscle	353.833
Stomach	288.520
Small Intestine	252.117
Colon	313.618

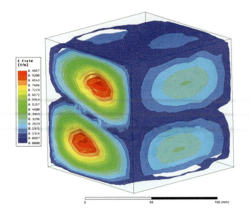

Figure 5.19 Simulated electric field variation inside the phantom box.

space wavelength at the operating frequency for avoiding an unwanted spurious mode.

This cube mimics muscle tissue, which has an electrical conductivity of 1.705 S/m and permittivity of 52.79 at 2450 MHz. The designed structure offers an impedance bandwidth of 630 MHz (2160–2940 MHz), which is 30.60% fractional bandwidth with respect to its center frequency, as shown in Figure 5.20.

The variation of the real and imaginary parts of the antenna's impedance with regard to frequency is shown in Figure 5.21. It has been observed that the real part reaches almost 50 Ω at 2450 MHz, while the imaginary part reaches zero, which suffices for design purposes.

The interior electronic components of capsule can also lower the performance of the antenna. The battery module especially plays a very important role. So this phenomenon is studied with a cylindrical battery (considered a perfect electrical conductor). The performance parameters are shown in Figure 5.22. Apart from the frequency detuning effect, the peak gain is also reduced by 2–3 dB.

As the capsule traverses various biological media during the actual endoscopy operation, the antenna will always see its variable surroundings around it. This can lead to a change in frequency response as well as the radiation

Printed Antennas for Biomedical Applications 121

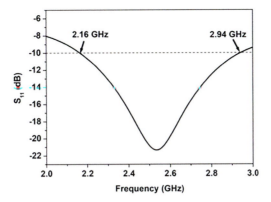

Figure 5.20 Simulated reflection coefficient of the capsule antenna at the center of the muscle-tissue phantom.

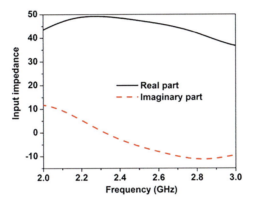

Figure 5.21 Variation of input impedance of proposed antenna with frequency.

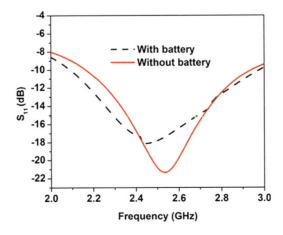

Figure 5.22 WCE antenna with and without battery inside the capsule.

pattern of the antenna. To analyze this effect in the simulation, various possible phantom conditions were created, which resulted in comparative performance data, as shown in Figure 5.23. Table 5.6 summarizes a data set for various biological tissues at 2450 MHz. this analysis shows that there is a slight decrease of the antenna's resonance frequency in the stomach and colon due to higher permittivity. Despite these variations, the reflection coefficient of the antenna remained below 10 dB between 2.2 and 2.6 GHz in all cases.

5.3.2 Electrical Equivalent Circuit

The whole antenna assembly can be modeled as a superstrate–substrate composite structure, as shown in Figure 5.24(a). Here, body tissue acts as a superstrate. This can also be expressed using a transmission line analogy, as explained in Figure 5.24(b).

Similarly, with the help of the lumped R, L, and C parameters, the electrical equivalent circuit of the antenna can be envisioned as shown in Figure 5.25.

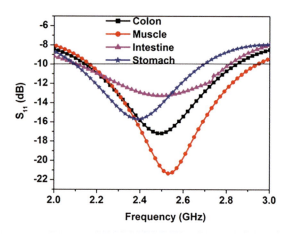

Figure 5.23 Simulated reflection coefficient of WCE antenna for different tissues of GI tract.

Table 5.6 Dielectric Properties of Different Biological Tissues at 2.45 GHz

Tissues	ε_r (Permittivity)	σ Conductivity) (S/m)	Gain (dBi)	BW (MHz)
Muscle	52.79	1.705	−30.61	630
Stomach	62.239	2.1671	−33.81	580
Small Intestine	54.527	3.1335	−45	690
Colon	53.969	1.9997	−33.63	680

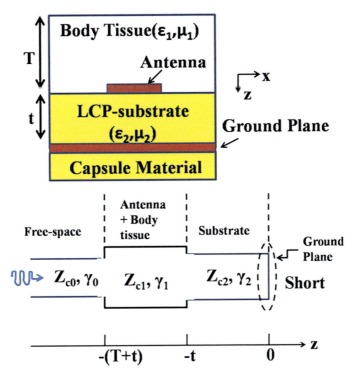

Figure 5.24 (a) Endoscopy antenna embedded in composite structure. (b) Transmission line analogy of superstrate–substrate geometry.

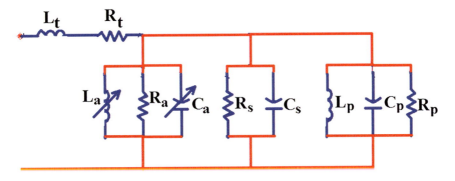

Figure 5.25 Electrical equivalent circuit.

Phantom liquid exhibits electrical conductivity as well as a dielectric nature and surrounds the capsule antenna.

Here a cubical phantom is taken into consideration. Each side of the cube is represented by Z_p Ohm. Z_p is a parallel combination of an inductor and leaky capacitor. The dielectric property of the phantom is represented by C_p, while the finite conductivity of bio-fluid is represented by L_p and R_p. The phantom environment is electromagnetically coupled with the capsule antenna by the flux linkage between L_p and L_a (inductance of the antenna). Thus, the environment inductance changes dynamically depending upon the nature of the bio-tissue. It is reflected in the S_{11} curve by the detuning factors observed thereafter. The antenna is represented by a tank circuit (parallel RLC network), tuning at TM_{10} mode. Effective inductance and capacitance determine the center frequency, whereas the resistance changes its impedance bandwidth profile. Little electromagnetic energy is wasted within the substrate material in terms of leaky and surface waves. It happens due to the non-zero value of the loss tangent parameter, pinholes associated with the dielectric material, or trap charges induced during manufacture of the substrate material. Substrate loss can be explained with the help of a leaky capacitor concept, that is, R_s in parallel with C_s. The feed portion of the antenna is expressed in terms of the series combination of R_t and L_t. This indicates the finite conductivity of the metal used in the feed line. For a certain reference case (muscle phantom), numerical values of each circuit parameter can be determined by the following mathematical equations, and Table 5.7 summarizes all obtained parameter values.

$$Z_{ocp} = \frac{30\pi}{\sqrt{\varepsilon_e}} \times \frac{K'(k)}{K(k)} \tag{5}$$

where

$$k = \frac{a}{b}\sqrt{\frac{1 - \left(\frac{b^2}{c^2}\right)}{1 - \left(\frac{a^2}{c^2}\right)}}, K'(k) = k\left(\sqrt{1 - k^2}\right)$$

Table 5.7 Numeric Values of Equivalent Circuit Parameters

Parameters	Values
R_t	24 Ω
L_t	1.21 nH
L_a	0.37 nH
R_a	43 Ω
C_a	0.21 pF
R_s	1000 Ω
C_s	32.34 pF
L_p	0.2 nH
C_p	1 pF
R_p	100

Printed Antennas for Biomedical Applications 125

Table 5.8 Maximum Power Limit for Satisfying the Standard SAR in the Human Body at 2.45 GHz

Antenna Locations	1-g SAR Power Limit (mW)
Muscle	3.2
Stomach	9.5
Colon	8
Small intestine	6.3

$a = W_f/2$; $b = a + g_f$; $c = g_f + 2.5\ W_f$
ε_e = effective permittivity of the substrate
$K(k)$ = complete elliptical integral of the first kind
k = aspect ratio
Like the previous capsule antenna case, this antenna also suits the bio-compatible features in terms of SAR value. For different bio-tissues, it has been evaluated, as shown in Table 5.8.

5.3.3 Measurement of Antennas

After fabricating the prototype antenna, the substrate was wrapped over the outer wall of the endoscopy capsule. Afterwards, a 0.1-mm-thick spray coating of PEEK (ε_r = 3.2, tanδ = 0.01) layer was applied, which is bio-compatible. Further, a suitable RF connector was assembled to the antenna circuit, as shown in Figure 5.26. In vitro measurement was done by a VNA, R&S-ZVA40, and a cylindrical beaker to form the human phantom, as depicted in Figure 5.27.

The beaker was filled with a liquid mixture mimicking muscle tissue at 2450 MHz. The liquid was made by mixing 26.7% (by volume) diethylene glycol butyl ether (DGBE: $C_8H_8O_3$) with 73.2% deionized (DI) water and 0.04% salt (NaCl). The salted aqueous solution was prepared with solid contents of 5 gm/L. It is an organic compound, which has a very high boiling point. Figure 5.28 shows a capacitive result between measured and simulated reflection coefficient data for the capsule antenna. The difference is attributed to fabrication tolerances and unavoidable human errors associated with the preparation of the phantom liquid mixture.

As explained in the previous section, here also it is necessary to evaluate the characteristics of the human phantom very well because the dielectric loading effect can alter the whole resonance behavior of the antenna.

The far-field simulated radiation pattern of the capsule antenna with all its expected surroundings is shown in Figure 5.29. The E plane is almost like a figure eight, and the H plane is nearly omnidirectional in nature. A slight detrimental effect is observed in the presence of the battery/cell.

Figure 5.26 Fabricated antenna wrapped outside the capsule.

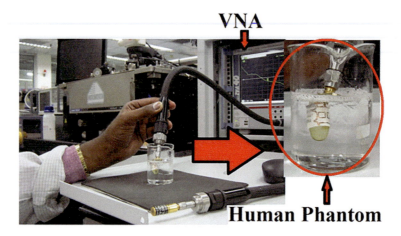

Figure 5.27 Measurement setup for fabricated antenna.

5.4 LINK BUDGET ANALYSIS

To predict the range of data telemetry, a link-budget analysis was computed. Figure 5.30 shows a schematic setup for link-budget calculation. The proposed antenna (either inside or outside the capsule) was considered a transmitting antenna, and a half-wave short dipole was considered a received one. Polarization and impedance matching loss were neglected. Under far field conditions, the link margin (LM) using the Friis transmission formula can be calculated as follows in Equation (6).

$$\begin{aligned} LM \text{ (dB)} &= \text{Link } C/N_0 - \text{Required } C/N_0 \\ &= \left(P_t + G_t + G_r - L_f - N_0\right) - \left(E_b / N_0 + G_d\right) \end{aligned} \tag{6}$$

Printed Antennas for Biomedical Applications 127

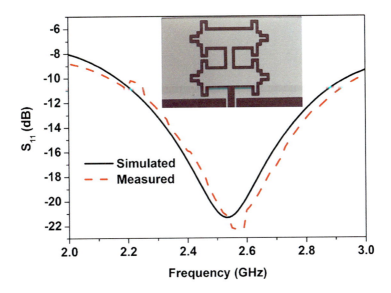

Figure 5.28 Simulated and measured reflection coefficient of proposed conformal antenna.

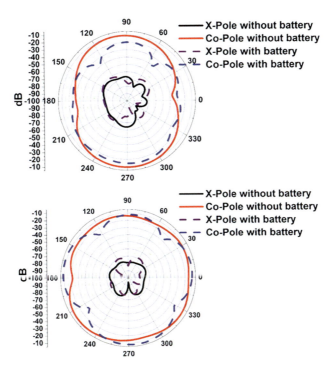

Figure 5.29 Radiation pattern characteristics of the proposed antenna with and without battery. (a) E-plane. (b) H-plane at 2.45 GHz.

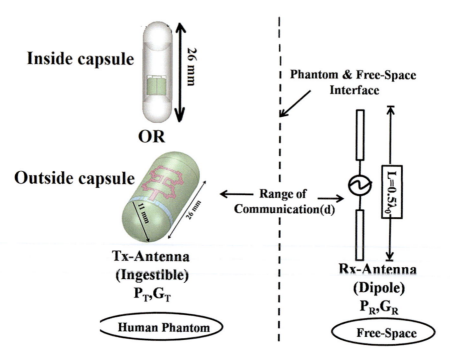

Figure 5.30 Schematic setup for link-budget analysis.

Table 5.9 Parameters of Link Margin Calculation

Parameters	Inside Capsule	Outside Capsule
Transmission		
Frequency (MHz)	915	2450
Tx power (P_t) [dBm]	−10	−4
Tx antenna gain (G_t) [dBi]	−35.5	−30.61
EIRP = $P_t G_t$ [dBm]	−45.5	−34.61
Propagation		
Distance d (m)	1, 5, 10	1, 3, 5, 7
Free space loss (dB)	31.66, 45.64, 51.66	40.257, 49.799, 54.236, 57.158
Receiver-Dipole Antenna		
Rx antenna gain G_r (dBi)	2.15	2.15
Temperature T_0 (Kelvin)	293	293
Boltzmann constant (J/K)	1.38×10^{-23}	1.38×10^{-23}
Noise power density N_0 (dB/Hz)	−200	−199.95
Signal Quality		
E_b/N_0 (ideal PSK) (dB)	9.6	9.6
Bit error rate (P_e)	1×10^{-5}	1×10^{-3}
Bit-rate (Mbps) (B_r)	1	1

Parameters	Inside Capsule	Outside Capsule
Coding gain (G_c)	0	0
Fixed deterioration G_d (dB/Hz)	2.5	2.5
Link Margin		
Distance d (m)	1, 5, 10	1, 3, 5, 7
LM (dB)	52.84, 38.86, 32.84	48.144, 38.60, 34.165, 31.243

where P_t = input power of capsule antenna, G_t = gain of capsule antenna, G_r = gain of receiver antenna, L_f = free space path loss [18], N_0 = noise power spectral density, E_b = bit error ratio, G_d = different deterioration, and C/N_0 = carrier-to-noise ratio.

Further,

$$L_f = 10\log_{10}\left(\frac{4\pi d}{\lambda_0}\right)^2$$

where d = reference distance from Tx to Rx antenna

According to [19], the acceptable path loss for an improved WCE was 73.3 dB.

5.5 SUMMARY

In this chapter, two types of printed antennas are discussed for biomedical applications in the ISM band. One miniaturized antenna was placed inside a capsule with an operating frequency of 915 MHz, and another antenna was wrapped outside the capsule wall with an operating frequency of 2.45 GHz. Parametric studies of the antennas were done to obtain the optimum design values for both antennas. The maximum simulated gain the inside antenna was −35.5 dBi in the colon phantom, whereas it was −30.61 dBi for the muscle phantom in the conformal antenna. Both antennas were fabricated and measured in two different substrates, high-resistive silicon and liquid crystal polymer, respectively. The variation of antenna performance with different tissues was also studied, and it was shown that both antennas had satisfactory performance in various tissues due to their wide bandwidth.

REFERENCES

[1] D. Nikolayev, M. Zhadobov, L. L. Coq, P. Karban, and R. Sauleau, "Robust Ultraminiature Capsule Antenna for Ingestible and Implantable Applications," *IEEE Transactions on Antennas and Propagation*, Vol. 65, no. 11, pp. 6107–6119, 2017, doi: 10.1109/TAP. 2017.2755764.

[2] C. Liu, Y. X. Guo, and S. Xiao, "Circularly Polarized Helical Antenna for ISM-Band Ingestible Capsule Endoscope Systems," *IEEE Transactions on Antennas and Propagation*, Vol. 62, no. 12, pp. 6027–6038, 2014.

[3] G. J. Gonzalez, D. Sadowski, K. Kaler, and M. Mintchev, "Ingestible Capsule for Impedance and PH Monitoring in the Esophagus," *IEEE Transactions on Antennas and Propagation*, Vol. 54, no. 12, pp. 2231–2236, 2007.

[4] H. Jacob, D. Levy, R. Shreiber et al., "Localization of the Given M2A Ingestible Capsule in the Given Diagnostic Imaging System," *Gastrointestinal Endoscopy*, Vol. 55, p. 135, 2002.

[5] J. Kim, and Y. Rahmat-Samii, "Implanted Antennas Inside a Human Body: Simulations, Designs and Characterizations," *IEEE Transactions on Microwave Theory and Techniques*, Vol. 52, no. 8, pp. 1934–1943, 2004.

[6] A. Kiourti, K. A. Psathas, and K. S. Nikita, "Implantable and Ingestible Medical Devices with Wireless Telemetry Functionalities: A Review of Current Status and Challenges," *Bioelectromagnetics*, Vol. 35, no. 1, pp. 1–15, 2015.

[7] S. H. Lee et al., "A Wideband Spiral Antenna for Ingestible Capsule Endoscope Systems: Experimental Results in a Human Phantom and a Pig," *IEEE Transactions on Biomedical Engineering*, Vol. 58, no. 6, pp. 1734–1741, 2011.

[8] H. Rajagopalon, and Y. Rahmat-Samii, "Wireless Medical Telemetry Characterization for Ingestible Capsule Antenna Designs," *IEEE Antennas Wireless Propagation Letters*, Vol. 11, pp. 1679–1682, 2012.

[9] R. Das, and H. Yoo, "A Wideband Circularly Polarized Conformal Endoscopic Antenna System for High-Speed Data Transfer," *IEEE Transactions on Antennas Propagation*, Vol. 65, no. 6, pp. 2816–2826, 2017.

[10] S. R. Gurudu, H. E. Vargas, and J. A. Leighton, "New Frontier in Small-Bowel Imaging: The Expanding Technology of Capsule Endoscopy and Its Impact in Clinical Gastroenterology," *Reviews in Gastroenterological Disorders*, Vol. 8, pp. 1–14, 2008.

[11] R. Garg, K. C. Gupta, and I. J. Bahl, *Microstrip Lines and Slotlines* (1st edition), Norwood: Artech House, ISBN-10: 0890060746, 1979.

[12] C. Liu, Y. X. Guo, and S. Xiao, "Capacitively Loaded Circularly Polarized Implantable Patch Antenna for ISM Band Biomedical Applications," *IEEE Transactions on Antennas Propagation*, Vol. 62, no. 5, pp. 2407–2417, 2014.

[13] C. Liu, Y. X. Guo, and S. Xiao, "Circularly Polarized Helical Antenna for ISM-Band Ingestible Capsule Endoscope System," *IEEE Transactions on Antennas Propagation*, Vol. 62, no. 12, pp. 6027–6039, 2014.

[14] N. Haga, K. Saito, M. Takahashi, and K. Ito, "Characteristic of Cavity Slot Antenna for Body-Area Networks," *IEEE Transactions on Antennas Propagation*, Vol. 57, no. 4, pp. 837–843, 2009.

[15] W. Xia, K. Saito, M. Takahaski, and K. Ito, "Performances of an Implemented Cavity Slot Antenna Embedded in the Human Arm," *IEEE Transactions on Antennas Propagation*, Vol. 57, no. 4, pp. 894–899, 2009.

[16] R. Warty, M. R. Tofighi, U. Kawoos, and A. Rosen, "Characterization of Implantable Antennas for Intracranial Pressure Monitoring Reflection by and Transmission through a Scalp Phantom," *IEEE Transactions on Microwave Theory and Techniques*, Vol. 56, no. 10, pp. 2366–2376, 2008.

[17] U. Kawoos, M. R. Tofighi, R. Warty, F. A. Kralick, and A. Rosen, "In-Vitro and In-Vivo Trans-Scalp Evolution of an Intracranial Pressure Implant at 2.4 GHz," *IEEE Transactions on Microwave Theory and Techniques*, Vol. 56, no. 10, pp. 2356–2365, 2008.

[18] W. Xia, K. Saito, M. Takahashi, and K. Ito, "Performance of an Implanted Cavity Slot Antenna Embedded in the Human Arm," *IEEE Transactions on Antennas Propagation*, Vol. 57, no. 4, pp. 894–899, 2009.

[19] K. Takizawa, H. Hagiwara, and K. Hamaguchi, "Design of a Radio System for Capsule Endoscopy Through a Path Loss Analysis," *6th European Conference on Antennas and Propagation* (EUCAP 2012), Prague, Czech Republic, pp. 553–556, March 2012.

Problems

1. What is SAR in biomedical engineering?
2. What is the significance of wireless capsule endoscopy (WCE)?
3. How do antennas play an important role in WCE systems?
4. What is bio-compatible material?
5. What is link-budget analysis? Why is it essential?
6. What is a link margin?
7. How does a battery play an important role in antenna design?
8. How do you characterize WCE antennas?
9. What are the different challenges in designing antennas for biomedical applications?
10. Why is an omnidirectional pattern essential for WCE antennas?

Chapter 6

High-Gain Printed Antennas for Sub-Millimeter Wave Applications

6.1 INTRODUCTION

Printed antennas for the sub-millimeter wave range are obviously miniaturized in shape. But, apart from their compact size, they also require a high-gain profile, which is traditionally achieved with bulky mechanical structures. Whenever, we think about any system-on-chip (SoC) concept at millimeter wave (mmW) or sub-mmW frequency range, designing the antenna also plays a very important role. In the field of medical imaging or the area of Terahertz application, antennas with high gain and wide band characteristics are especially essential. This chapter discusses printed versions of miniaturized antennas with high-gain profiles targeting sub-mmW applications.

6.2 DESIGN AND DEVELOPMENT OF HIGH-GAIN PRINTED ANTENNAS

This section deals with the design and development of a high-gain miniaturized antenna for sub-millimeter wave application. It is one kind of antenna array where parasitic elements have been loaded to enrich the peak gain. The antenna structure is fabricated on a thin (~ 4 mil), flexible substrate called liquid crystal polymer. This substrate is a non-toxic material and bio-compatible in nature. The literature shows that RF circuits have been demonstrated on LCP even up to 110 GHz [1–5]. This material is compatible with the standard CMOS process or micro-fabrication method. In the current work, we have chosen 4-mil-thick Rogers-made ULTRALAM 3850-HT LCP substrate, properties of which are summarized in Table 6.1.

The details of antenna engineering are given in the following.

DOI: 10.1201/9781003389859-6

134 Printed Antennas for Wireless Communication and Healthcare

Table 6.1 Properties of ULTRALAM-3850 HT LCP Substrate

Properties	Typical Values
Dielectric constant	2.92–3.32
Dissipation factor	0.0020–0.0038
Thermal Coefficient (ε_r)	9 ppm/$^\circ$C (−100°C to +250°C)
Density	1.4 gm/c.c
Dielectric strength	2.27 kV/mil
Volume resistivity	10^{10} MΩ-cm
Moisture absorption	<0.4%
Thermal conductivity	0.3 W/m/k

6.2.1 Design of Parasitic Element Loaded Microstrip Antenna Arrays With Enriched Gain

This section details the design of a rectangular microstrip antenna array (RMAA) loaded with parasitic elements to achieve very high gain. The targeted frequency of application is 100 GHz. A high peak gain of around 19.3 dBi has been achieved in this planar configuration with five radiating elements connected in cascade with series feeding. Each of the radiating elements excites two parasitic patches placed on both sides of the non-radiating edges. Figure 6.1 shows the schematic of the antenna array. This array architecture comprises five primary patches (length L and width W) with ten parasitic elements (length L and width Wp). The dimensions of the patch can be evaluated by Equations (1)–(2) [6].

$$W = \frac{C}{2f_r} \sqrt{\frac{2}{1 + \varepsilon_r}} \qquad (1)$$

$$L = \frac{C}{2\sqrt{\varepsilon_{eff}}} \frac{1}{f_r} - 2\Delta L \qquad (2)$$

where
f_r = 100 GHz
ε_{eff} = elective dielectric constant
ΔL = extended length on both sides of the radiating patch due to fringing field [6]

In this series-fed array configuration, small segments of transmission line between primary patches are used for impedance matching as well as for phase-balancing purposes. The dimensions of this Tx-line segment, length ℓ_{t1} and ℓ_{t2} and width W_{t1} and W_{t2}, are computed by standard design equations available in [7]. Parasitic elements, along with the primary radiators, enhance

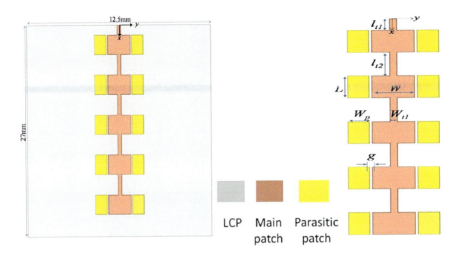

Figure 6.1 Top view and dimensions of RMAA with parasitic patches.

the aperture size of the array structure, which in turn becomes responsible for the high-gain profile.

Parametric study has been carried out to optimize various design parameters for this array design. In particular, here the gap between the main radiating elements and parasitic patch and the width of parasitic elements play important roles. Figure 6.2 depicts an abstract of the parametric study. The gap g determines the amount of E-field coupling strength between the main radiating element and the parasitic patches [8]. As the antenna is realized implementing the standard PCB etching chemistry, hence any design dimension below 3 to 4 mil is hard to achieve. Simulation study reveals that up to g values of 200 μm, the reflection coefficient of the antenna doesn't alter significantly.

While the width of the parasitic patch is changed, the overall input impedance of the antenna is changed, and other spurious modes also try to appear. The surface current distribution plot explains this phenomenon clearly. Finally, the optimized dimensions of the RMAA architecture are summarized in Table 6.2.

The surface current distribution of this enhanced gain array antenna is depicted in Figure 6.3. It reveals that the energy associated with each patch element monotonically decreases from the directly fed array element to the last element. This is because most of the energy is radiated by the first patch itself, and the rest is passed to the next patch, which radiates a partial amount of energy, partially transfers to the next patch, and so on.

The simulated far-field radiation pattern of the antenna is shown in Figures 6.4 and 6.5. The H-plane indicates a broadside pattern in the bore sight,

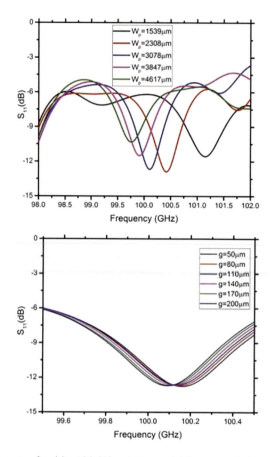

Figure 6.2 S parameter for (a) width W variation and (b) gap g variation.

Table 6.2 Optimized Dimensions of RMAA

Variables	Without Parasitic Elements (μm)	With Parasitic Elements (μm)
L	2489	2489
W	3078	3078
l_{t1}	1293	1293
l_{t2}	2685	2685
w_{t1}	690	500
w_p	–	1539
g	–	200
h	100	100

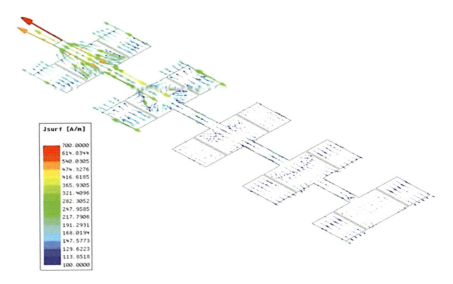

Figure 6.3 Surface current distribution for RMAA.

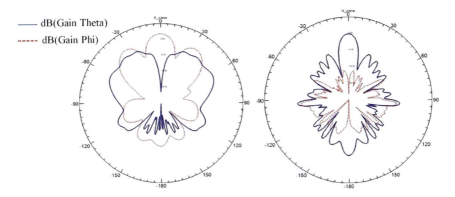

Figure 6.4 Simulated far-field radiation pattern of the RMAA at 100 GHz in (a) E-plane and (b) H-plane.

and the cross-polarization level is well below (~38 dB) the co-polarization level, whereas the E-plane pattern is like a flower with multiple petals. Peak simulated gain is around 19.2 dBi, with a radiation efficiency of more than 80%.

6.2.2 Antenna Fabrication

After optimization, the final prototype structures are fabricated on 4-mil-thick LCP (ε_r = 2.92, tanδ = 0.002) material, as shown in Figure 6.6. Standard wet etching chemistry is implemented to pattern the metal layer (18 μm

138 Printed Antennas for Wireless Communication and Healthcare

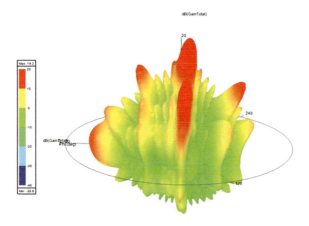

Figure 6.5 3D-radiation pattern of the RMAA with parasitic patches in H-plane.

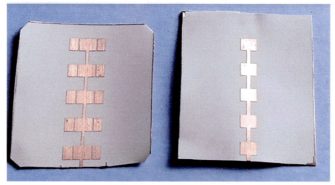

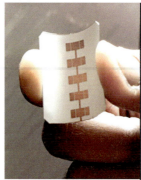

Figure 6.6 Fabricated prototype antenna on LCP: (a) array with/without parasitic elements; (b) the flexibility of the structure.

copper on the front side). To realize the microstrip ground plane, the backside metallization layer is kept untouched. A special type of RF-connector is essential for testing such antennas.

6.3 SUMMARY

This work focuses on exploring the capability of printed antennas with very high gain for sub-millimeter wave applications. Along with this, it also outlines the possibility of framing flexible electronics for such high-frequency usage. LCP is chosen for this R&D activity, and its flexibility, non-toxic characteristics, light weight, robustness in harsh environment, and above all ease of making circuits make this an attractive candidate for several commercial as well as strategic applications in the very near future. Prototypes have been fabricated for proof of concept.

REFERENCES

[1] N. Kingsley, "Liquid Crystal Polymer: Enabling Next-Generation Conformal and Multilayer Electronics," *Microwave Journal*, pp. 188–200, 2008.

[2] D. C. Thompson, O. Tantot, H. Jallageas, G. E. Ponchak, M. M. Tentzeris, and J. Papapolymerou, "Characterization of Liquid Crystal Polymer (LCP) Material and Transmission Lines on LCP substrates from 30 to 110 GHz," *IEEE Transaction on Microwave Theory and Techniques*, Vol. 52, no. 4, pp. 1343–1352, 2004.

[3] Georgia Tech Report, https://phys.org/news/2006-08-georgia-tech-liquid-crystal-polymer.html

[4] S. Horst, S. Bhattacharya, S. Johnston, M. M. Tentzeris, and J. Papapolymerou, "Modeling and Characterization of Thin Film Broadband Resistors on LCP for RF Applications," *Procs. of the 2006 IEEE-ECTC Symposium*, pp. 1751–1755 (San Diego), 2006.

[5] K. Sangkil, R. Amin, L. Vasileios, N. Symeon, M. M. Tentzeris, "77 GHz mm Wave Antenna Array on Liquid Crystal Polymer For Automotive Radar and RF Front End Module," *ETRI Journal*, Vol. 41, no. 2, pp. 262–269, 2019.

[6] A. B. Constantine, *Antenna Theory Analysis and Design* (3rd edition), Hoboken: John Wiley & Sons, Inc., 2005.

[7] M. S. Rabbani, and H. Ghafouri-Shiraz, "Improvement of Microstrip Antenna's Gain, Bandwidth and Fabrication Tolerance at Terahertz Frequency Bands," *Wideband and Multi-Band Antennas and Arrays for Civil, Security & Military Applications*, London, 2015, pp. 1–3, doi: 10.1049/ic.2015.0146.

[8] G. Kumar, and K. P. Ray, *Broadband Microstrip Antenna*, Norwood: Artech House Publisher Inc., pp. 113–114, 2003.

Problems

1. What is flexible electronics?
2. Give some examples of flexible electronics.
3. What is LCP?
4. Why is LCP substrate becoming popular for current electronics?
5. What are the different ways to improve the gain of printed antennas?
6. What is the sub-mmW frequency range? What are its usages?
7. Why is a high-gain antenna required for the sub-mmW band?
8. What is capacitive loading in antenna engineering?
9. What are the different features of substrates required for biomedical applications?
10. How are leaky-wave and surface-wave problems associated with antenna substrate height and its dielectric properties?

Chapter 7

Systematic Investigation of Various Common Imperfections in Printed Antenna Technology and Empirical Modeling

7.1 INTRODUCTION

In the last six chapters, the authors have discussed the evolution of printed antennas along with their wide application. Then we outlined various modeling approaches for those antenna modules in context with equivalent RLC networks. How miniaturization techniques are adopted for modern antennas has also been outlined. The use of fractal geometry in antenna engineering is also one booming technique. Apart from the conventional RF/wireless communication field, miniaturized antennas are finding great importance in the modern-day biomedical field. They are becoming an inevitable part of our day-to-day life. People have reported numerous techniques of printed antenna development work in the last 30 years, but every method has its own pros and cons.

In this chapter, we focus on some of the common imperfections involved in the development of printed antennas. In parallel, each phenomenon will be expressed with an empirical model, which sheds significant light on the problem for better understanding of the imperfection. Broadly, these imperfections are categorized into two classes: fabrication related and assembly related. The following sections will examine them.

7.2 FABRICATION-RELATED IMPERFECTIONS

This section will discuss various common fabrication issues of printed antennas. These are as follows.

7.2.1 Surface Roughness or Hillocks

In an active or passive device, the conduction current flows through the metallization layer, though theoretically it is assumed that the metal surface is ultra-smooth. In practice it doesn't happen this way because of fabrication-related irregularities or environmental effects, such as corrosion. In the case of RF devices, this roughness plays a very important role, especially at the

DOI: 10.1201/9781003389859-7

higher side of the EM spectrum, while the roughness profile is comparable to the skin depth. The conductivity of metal and its surface impedance can be changed significantly because of this imperfection. Usually two popular models are adopted to investigate surface roughness–related matters: the Hammerstad-Bekkadal [1, 2] and Huray [3–5] models.

7.2.1.1 Hammerstad-Bekkadal Model

In this model, surface roughness is treated as series of peaks and valleys, as shown in Figure 7.1(a). This model yields good results when the RMS value of roughness (h_{RMS}) is comparable with skin depth (δ_{skin}). The roughness factor (f_r) given in this model is expressed as

$$f_r = 1 + \frac{2}{\pi} \tan^{-1}\left[1.4\left(\frac{h_{RMS}}{\delta_{skin}}\right)^2\right] \tag{7.1}$$

where δ_{skin} is given by

$$\delta_{skin} = \left(\frac{2}{\sigma_{smooth}\omega\mu}\right)^{1/2} \tag{7.2}$$

Hence, the modified conductivity of material (σ_{rough}) is

$$\sigma_{rough} = \frac{\sigma_{smooth}}{f_r} \tag{7.3}$$

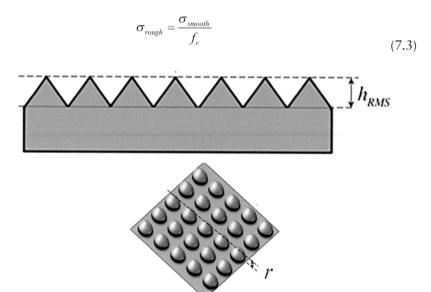

Figure 7.1 Roughness profiles of surface assumptions for the (a) Hammerstad-Bekkadal model and (b) Huray model.

Systematic Investigation of Various Common Imperfections   143

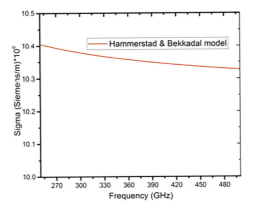

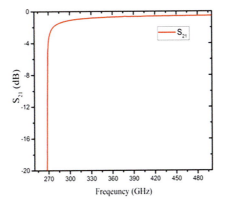

Figure 7.2 (a) Variation of conductivity with respect to frequency and (b) S_{21} plot for Hammerstad-Bekkadal model.

Equations (7.1)–(7.3) give an idea of the modified metal conductivity in context with surface roughness. It clearly indicates that the skin-depth can also be changed. The variation of metal conductivity with respect to frequency is shown in Figure 7.2(a), and how this modulated conductivity can affect the attenuation factor in the feed network of the antenna structure is also shown in Figure 7.2(b) using the Hammerstad-Bekkadal model.

7.2.1.2 Huray Model

This model considers the roughness profile through hemispherical snowballs of radius r on the surface, as shown in Figure 7.1 (b). The roughness factor (f_r) in this model is expressed as

$$f_{r_{Hall\text{-}Huary}} = 1 + \frac{3}{2} \frac{\frac{4\pi a^2 N}{A}}{1 + \frac{\delta_{skin}}{a} + \frac{\delta_{skin}^2}{2a^2}} \quad (7.4)$$

where, N is expressed as

$$N = \frac{Volume\ of\ surface}{Volume\ of\ each\ Snow\ ball\ (sphere)} \quad (7.5)$$

Considering a metallization thickness of 3 μm for the Huray model, the radius of the snowball can typically range between 0.1 and 1.5 μm. Figure 7.3 depicts the modified conductivity due to the surface roughness factor, along with the attenuation factor.

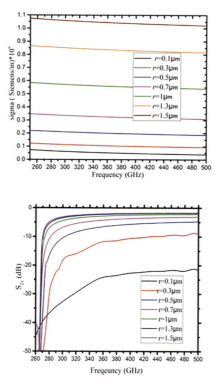

Figure 7.3 (a) Variation of conductivity with respect to frequency and b) S_{21} plot for different radii of snowballs using the Huray model.

7.2.2 Voids in Via-Hole Filling

Interconnections and grounding play a key role in advanced radio frequency and mixed-signal integrated circuits. Vias are implemented in single-layer or multilayer PCBs and RFICs to provide ground paths or connect various layers [8]. Printed antenna technology often uses this via-hole architecture to enhance the gain and power handling capability. The literature has reported prediction of an accurate, simple, and scalable model of the via-hole [9–10]. But there is hardly any literature which talks about the practical imperfections imposed by the fabrication process on via-hole modeling. Generally, it is a very common issue of the formation of voids with the plating (or metallization) of the via-hole structure. This can lead to a complete change of the current distribution profile. The whole empirical circuit model can be now thought of as shown in Figure 7.4.

Overall inductance, capacitance, and resistance profiles are altered significantly with the inclusion of voids in the metallization pattern. This change in values of R_{via}, L_{via}, and C_{via} depends upon the size and number of voids formed. There may even be formation of an extra parasitic capacitance (C_{para}) due to this.

7.2.3 Voids in Traditional Conducting Plates of Radiator or Feed Networks

Due to unexpected contaminations present in the pre-metal process step(s), there may be a chance of void formation in the metallization layer. This void can give rise to a change of current vector direction as well magnitude along with extra inclusions of fringing fields. Though this effect can be neglected in DC or in lower-frequency applications, while entering the field of microwave

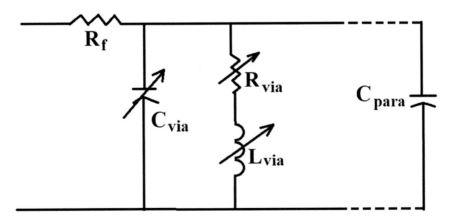

Figure 7.4 Proposed electrical equivalent of via-hole with void in its plating.

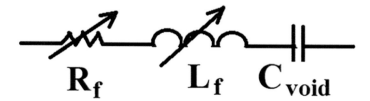

Figure 7.5 Electrical equivalent of faulty metalized feed structure with voids or scratches.

or millimeter waves or a sub-mmW regime, huge signal loss or a change in the current vector can deteriorate the expected results. A simple metal line with a void structure can then be modeled as a series of R_f, L_f, and C_{void}, as shown in Figure 7.5, where R_f and L_f are variable in nature, depending upon the nature of void formation.

7.2.4 Unexpected Polymer Residues in the Gap Between Two Metal Traces

Though proper cleaning steps are included within micro-fabrication processes after metal etching (whether dry or wet), there may a small chance of polymer residue remaining if the gap between two consecutive metal strips is very narrow. There is a probability of such an incident in the case of very high frequency RF-ICs or a system-on-chip (SoC), where metal traces are densely populated. In an antenna array structure, various metallic lines run in parallel side by side in close proximity. These unexpected residues may sometimes short two strips or invite perturbation in the electric field between two adjacent traces. This imperfection may be electrically modeled as follows (Figure 7.6). Depending upon the volume and position of the polymer residue, the value of $C_{residue}$ changes.

7.2.5 Cavity Opening Problem in Bulk Micro-Machined Antennas

Bulk micromachining is a process which is widely used for realization of multiple kinds of MEMS antennas. Whether it is partial etching or through and through wafer etching [6, 7], it plays a key role in establishing the merits of a MEMS antenna over its traditional printed version counterparts. Usually, in an anisotropic wet etching method with the help of tetra methyl amino hydroxide (TMAH) and potassium hydroxide (KOH), silicon wafer micromachining is performed. Between these two, KOH has become more popular because of its inherent advantages of a less toxic nature and high etch-rate of 1.1–1.2 μm/min, compared to 0.8 μm/min for TMAH solution. A standard process of 40% KOH solution at 80°C is the optimum method for silicon micromachining, but it requires continuous stirring to keep the whole

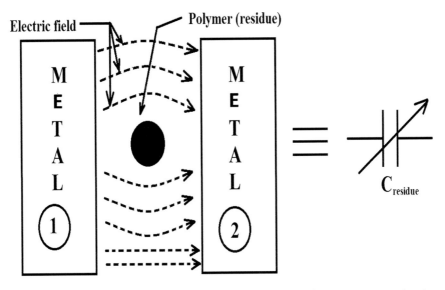

Figure 7.6 Schematic model of the polymer residue remaining between two metal strips in a high-frequency circuit.

solution homogeneous. Using the etch-stop method at the right instant controls this process. Although precautions [11, 12] are taken during processing, there may be a high chance of non-uniformity in the case membrane thickness at the last stage of the whole process. This non-uniformity may be of two types: within wafer or wafer-to-wafer. Now, suppose there is an antenna array that has N elements. Though during design stage, the antenna was optimized by assuming uniform membrane thickness beneath each element, however due to variation in the fabrication processes, each of the radiating element looks at a non-uniform silicon membrane below it. Ultimately the resonant frequency will be changed along with its other radiation characteristics. The whole phenomenon can be explained with the help of the proposed electrical model, as shown in Figure 7.7.

The center frequency of the antenna is controlled by the C_{10} and L_{10} values, while the impedance bandwidth profile is governed by R_{10}. For the non-uniform nature of the bulk micromachining process, the C_{10} value changes. C_{10} is the series equivalent capacitance of C_{ox} (capacitance due to the etch-stop SiO$_2$ layer), C_{Si} (capacitance due to the Si membrane), and C_{air} (capacitance due to air pocket formed between the Si wafer and bottom ground plate). Due to non-uniformity in etching, C_{Si} and C_{air} change dynamically, which changes the C_{10} values and in turn alters the resonant frequency $\left(f_0 = \dfrac{1}{2\pi\sqrt{L_{10}C_{10}}} \right)$ of the antenna.

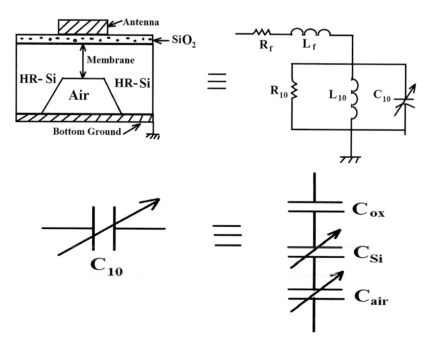

Figure 7.7 Proposed circuit model of (a) single element micro-machined patch antenna; (b) equivalent capacitance of the tank circuit showing its all components.

7.2.6 Sagging Problems in Surface Micromachining

In the realization of various tunable or reconfigurable antennas in RF-MEMS technology, usually a cantilever or fixed-beam configuration type of RF switch or its derived circuitries are employed. Surface micromachining is the backbone of realization of such devices and suffers mainly due to stiction. Many methods have been reported in recent times to alleviate this process issue [13]. But it is really very difficult to get a perfectly straight hanging membrane without any sagging due to processing issues or internal residual stress problems.

Due to this sagging, an unexpected variation of the initial air gap-height occurs, which in turn can change the ON-state capacitance (C_{up}-state) value at the higher side. This way, the figure of merit of a capacitive structure deteriorates, which in turn leads to poor RF performances. The electrical equivalent of such an imperfection is shown in Figure 7.8.

7.2.7 Improper Wafer Bonding or Misalignment in 3D Vertical Integration

Wafer bonding is a unique feature of MEMS technology, which is not possible in the standard CMOS process. Various three-dimensional structures (cavities, resonators, filters, waveguides, aperture-coupled antennas,

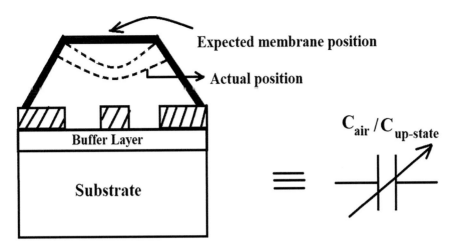

Figure 7.8 Electrical equivalent of sagging-induced membrane in RF switch structure.

antennas with multiple substrates, etc.) can be made faithfully in RF-MEMS with the aid of a bonding mechanism. Today, literature can be found where it has been shown that vertical integration is an inevitable technique which is often used in the realization of aperture-coupled antennas or antennas with multiple other building blocks in the system-on-chip concept. Usually, the anodic and eutectic categories are very popular [14, 15]. The main promising feature of this technique is that user can get a higher Q-factor–based cavity or resonator structure or an antenna with the desired air pocket beneath the patch per design requirements, which is hardly possible in the conventional microfabrication process because of several constraints. This whole process is governed by the ultra-clean surface of the bonded materials, and alignment accuracy is maintained throughout the process. A particle in the bonding surface can be the cause of a future void or crack and thus lead to leakage of the cavity. On the other hand, a misalignment can alter the shape of the intended cavity, and thus its capacity to store electromagnetic energy or Q-factor will be changed. Thus the eigenmode pattern will also be altered, which will dramatically vary the frequency characteristics. For a better understanding of this process-related error, in Figure 7.9, one cavity structure is shown. Here, it is observed that using the "split-block" technique, two similar halves of the devices are made separately, like for a rectangular cavity resonator. For the misalignment issue, the equivalent capacitance and inductance change because of the changes in the internal magnetic field distribution and electric field pattern. Hence, the whole frequency response of the circuit is changed. In the case of aperture-coupled antennas, a simple misalignment can be the cause of spurious radiation or inefficient excitation of the main radiating element.

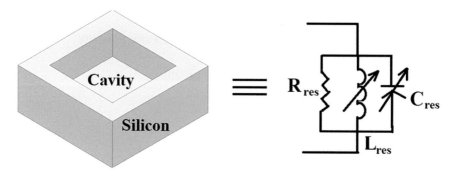

Figure 7.9 Proposed electrical equivalent of the cavity resonator structure.

7.3 IC ASSEMBLY OR PACKAGING-RELATED ISSUES

7.3.1 Effect of Epoxy Spreading

During assembly of the RF connector at the feed portion of the printed antenna, unexpected spreading of conductive epoxy or solder materials may happen, as the whole process is done manually. If it is a CPW-type feeding, then it can lead to a change in actual gap dimensions. Hence, the expected 50-Ω value of the standard characteristic impedance degrades, which may randomly affect the impedance matching profile. Simulations results show that the characteristic impedance can even fall in the range of 30 to 40 Ω instead of 50 Ω due to this.

The effect of epoxy has been analyzed in electro-magnetic solvers, as shown in Figure 7.10. It reveals that the S_{11} value can degrade up to 8 dB for the asymmetrical spreading of conductive epoxy at the feed portion of a simple 50-Ω CPW transmission line [16]. This imperfection in the IC assembly process can be electrically modeled as follows in Figure 7.11.

Due to unexpected spreading of conductive epoxy, with the transmission path, there is a certain localized change in capacitance (ΔC) and conductance (ΔG) values.

7.3.2 Improper Wire Bonding

Traditionally, the wire-bonding technology is used in IC packaging field to make out the electrical connection between chip-to-package. In System-in-Package (SiP) or Antenna-in-Package (AiP) during the realization of multichip modules (MCMs), wire bonds are absolutely essential. At the higher frequency range, they have a detrimental effect on RF performance. It can be thought of as a parasitic inductor series with a resistor if its

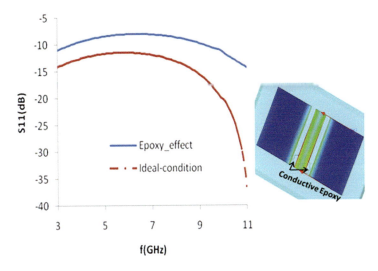

Figure 7.10 Effect of epoxy/solder spreading at the CPW-type feeding point of a printed antenna.

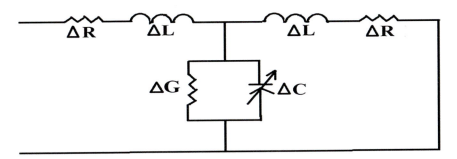

Figure 7.11 Proposed electrical equivalent of transmission line model with unwanted epoxy in the structure.

dimension is shorter than the operating wavelength. When a RF-current cycle increases and decreases in a sinusoidal fashion, the magnetic field surrounding a bond-wire expands and contracts, including a time-varying voltage. Ideally, bond-wires should be placed one-quarter wavelength (operating frequency) apart or less, not in an arbitrary fashion. Otherwise, their interactive coupling fields may affect the performance of the device. The skin-depth phenomena should also be taken into account. So the dimensions of the wire become very important. As a rule of thumb, a 1-mm-long, 1-mil wire provides 1 nH inductance [17]. However, in practice, people used to keep unnecessarily long bond-wires for their RF

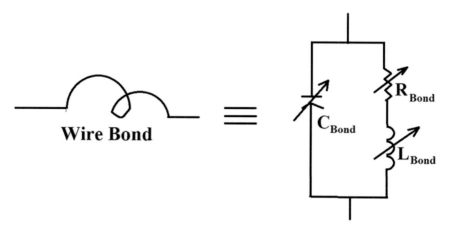

Figure 7.12 Proposed electrical equivalent of bond-wire geometry.

circuits or in some instances the bond wire is not terminated suitably. It will be a significant reason for signal loss and can even be a root cause of device failure. For 40–50 GHz or beyond, a wire bond may act like an electrically small antenna.

The proposed model (Figure 7.12) shows a simple bond-wire of improper detention or end connection can act as a tank circuit with variable frequency characteristics. With the change of basic bond-wire geometry, its parasitic inductance and capacitances are changed.

7.4 SUMMARY

This chapter discusses various process issues and packaging-related imperfections in the realization of printed antennas. Additionally, it explains every phenomenon with an electrical equivalent circuit model (empirical type). Thus, the working principle of the concerned antenna can be understood very easily. A total of 11 such practical scenarios have been demonstrated in this chapter. Per the authors' knowledge, such an attempt to explain all possible imperfections in printed antenna technology has not been tried to date. Failure analysis of the device can be done very easily using this work.

In the future, much more extensive research work has to be done in this field to predict the value of each circuit parameter. The vector-fitting curve method, along with experiments, can solve the mystery in a proper way.

REFERENCES

[1] B. Huang, and Q. Jia, "Accurate Modeling of Conductor Rough Surfaces in Waveguide Devices," *Electronics*, Vol. 8, pp. 1–12, 2019.

[2] E. O. Hammerstad, and F. Bekkadal, *Microstrip Handbook* (1st edition), Trondheim: Norwegian Institute of Technology, pp. 4–8, 1975.

[3] P. G. Huray, O. Oluwafemi, J. Loyer, E. Bogatin, and X. Ye, "Impact of Copper Surface Texture on Loss: A Model That Works," in *Proceedings of the Design Conference*, Santa Clara, pp. 462–483, 2010, https://www.oldfriend.url.tw/article/IEEE_paper/roughness/5_TA2_Paul_Huray.pdf.

[4] P. G. Huray, *The Foundations of Signal Integrity* (1st edition), Hoboken: John Wiley & Sons, Inc., pp. 216–261, 2009.

[5] P. G. Huray, S. Hall, S. Pytel, F. Oluwafemi, R. Mellitz, D. Hua, and Y. Peng, "Fundamentals of a 3-D "snowball" Model for Surface Roughness Power Losses. SPI," *IEEE Workshop*, pp. 121–124, 2007.

[6] K. E. Bean, "Anisotropic Etching of Silicon," *IEEE Transactions on Electron Devices*, Vol. 25, pp. 1185–1193, 1978.

[7] R. L. Marty, B. Saadany, B. Mercier, O. Français, Y. Mita, and T. Bourouina, "Advanced Etching of Silicon Based on Deep Reactive Ion Etching for Silicon High Aspect Ratio Microstructures and Three-Dimensional Micro and Nano-structures," *Microelectronics Journal*, Vol. 36, pp. 673–677, 2005.

[8] E. Holtzman, *Essentials of RF & Microwave Grounding* (1st edition), Boston: Artech House, pp. 61–67, 2006.

[9] M. E. Goldfarb, and R. A. Pucel, "Modeling Via Hole Grounds in Microstrip," *IEEE Microwave and Guided Wave Letters*, Vol. 1, pp. 135–137, 1991.

[10] P. Kok, and D. D. Zutter, "Capacitance of a Circular Symmetric Model of a Via Hole Including Finite Ground Plane Thickness," *IEEE Transaction on Microwave Theory Techniques*, Vol. 39, pp. 1229–1234, 1991.

[11] L. Wallman, J. Bengtsson, N. Danielsen, and T. Laurell, "Electrochemical Etch-Stop Technique For Silicon Membranes with P-Type and N-Type Regions and Its Application to Neural Sieve Electrodes," *Journal of Micromechanics and Microengineering*, Vol. 12, 2002.

[12] A. Karmakar, B. Biswas, and A. K. Chauhan. "Investigation of Various Commonly Associated Imperfections in Radio-frequency Micro-Electro-Mechanical System Devices and its Empirical Modeling," *IETE Journal of Research*, 2021, doi: 10.1080/03772063.2021.1880342

[13] Z. Yapu, "Stiction and Anti-Stiction in MEMS and NEMS," *Acta Mech Sinica*, Vol. 19, pp. 1–10, 2003.

[14] E. W. Hsieh, C. H. Tsai, and W. C. Lin, "Anodic Bonding of Glass and Silicon Wafers with an Intermediate Silicon Nitride Film and Its Application to Batch Fabrication of SPM Tip Arrays," *Microelectronics Journal*, Vol. 36, pp. 678–682, 2005.

[15] L. Overmeyer, Y. Wang, and T. Wolfer, "Eutectic Bonding," in *The International Academy for Production Engineering. CIRP Encyclopedia of Production Engineering*, Berlin: Springer, 2014.

[16] A. Karmakar, and K. Singh, "Planar Monopole Ultra Wide Band Antenna on Silicon with Notched Characteristics," *International Journal of Computer Application*, Vol. 3, pp. 17–20, 2014.

[17] A. Karmakar, and K. Singh, "Full-Wave Analysis and Characterization of RF Package for MEMS Applications," *Microwave Review*, Vol. 22, pp. 17–22, 2016.

Problems

1. What is stiction? How it can affect device performance?
2. How does bond wire play a key role in RF design?
3. How do you model a bond wire for high-frequency circuits?
4. What is the misalignment issue in the wafer bonding procedure?
5. What are the common fabrication issues in printed antennas?
6. What are the various assembly-related issues in antenna circuits?
7. How do you model a bulk-micromachined antenna with an improper cavity opening?
8. How do you model the plated through via hole?
9. What is the importance of plated through via technology in the antenna field?
10. What is 3D integration? How is it relevant to the antenna field?

Chapter 8

Multiple Input and Multiple Output Antennas

8.1 INTRODUCTION

Modern wireless communication world demands high-gain, compact, portable devices supported with enhanced data throughput for multiband utilization. The demand for higher data rates with longer distance has been creating a revolution behind the development of multiple inputs–multiple outputs (MIMO) technology. These days, WiFi, long-term evolution (LTE), and many other radio, wireless, and RF technologies are using the MIMO technique to acquire enhanced link capacity and spectral efficiency combined with improved link reliability. Work in this area started back in 1998; however, wide spreading of its applications/usages are observed in recent times because of its several promising features. In general, wireless channels may be affected by fading factors, thus leading to changes in the overall signal-to-noise ratio. In turn, this may impact the error rate. The concept of diversity is conceived to mitigate these kinds of losses.

There are three different diversity modes available for wireless communication systems: time diversity, frequency diversity, and space diversity. Using time diversity, a message may be transmitted at different time instants, whereas in frequency diversity, different frequency bands are used for operation. It may be by using different channels or technologies such as spread spectrum or OFDM. Space diversity employs multiple antennas located in different positions to take advantage of different radio paths. In a broader sense, this is the basis of MIMO technology. It can be viewed as a logical extension to the smart antennas that have been used for many years to improve the quality of wireless channels.

A general outline of a MIMO system is shown in Figure 8.1, where $[H]$ is the channel information matrix and h_{ij} is the fading coefficient of the ith transmitting and jth receiving antennas. Traditional MIMO systems may have two or four antennas at either or both the transmitter and receiver sides.

However, while tens or hundreds of antennas are used in MIMO systems, this is popularly termed massive MIMO, which creates more degrees

DOI: 10.1201/9781003389859-8

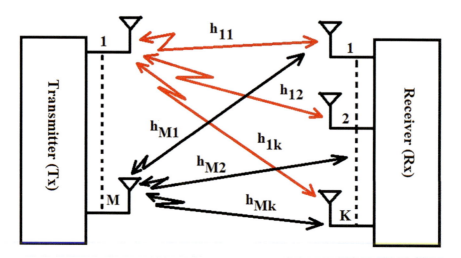

Figure 8.1 MIMO antenna arrangement.

of freedom in the spatial domain and therefore enables the whole system to improve.

Three basic performance metrics are enhanced with massive MIMO technology: increasing data rate, empowering the SNR, and hardening the wireless channel.

Depending upon the number of users in the MIMO system, it can be broadly classified into two categories: single-user MIMO (SU-MIMO) and multi-user MIMO (MU-MIMO).

For years, communication engineers assumed that the theoretical channel capacity limits were defined by the Shannon-Hartley theorem defined in Equation (1)

$$Capacity(C) = BW \times log_2[1 + SNR] \tag{1}$$

It demonstrates that an increase in a channel's SNR results in marginal gains in channel throughput. Hence, the traditional way to achieve higher a data rate is by increasing signal bandwidth. However, increasing the bandwidth of a communication channel by enhancing the symbol rate of a modulated carrier increases its susceptibility to multipath fading. A MIMO system offers an interesting solution to this fading issue by providing multiple antennas and multiple signal paths to gain knowledge of the communication channel. By using the spatial dimension of a communication link, a MIMO system can achieve significantly higher data rates than traditional single input–signal output (SISO) channel.

Here, the channel capacity can be estimated as a function of N spatial streams. Hence, Equation (1) can be modified as

$$Capacity = N \times BW \times log_2[1 + SNR] \tag{2}$$

Microstrip antennas (MSAs) are widely used in MIMO techniques due to their different advantages such as low cost, low profile, ease of fabrication, ease of integration with other microwave monolithic integrated circuits (MMICs), planar structure, and ability to generate both linear and circular polarization, though there are also different disadvantages to MSAs such as low gain, narrow bandwidth, low efficiency, and lower power handing capacity. Generally microstrip antennas can be rectangular, circular, square, elliptical, and triangular in shape, but in MIMO design, rectangular and circular antennas are preferable because of the characteristics of low cross-polarization level and low side-lobe level. If the length L and width W of a rectangular antenna are used for the MIMO technique, then W can be calculated by Equation (3) [1]

$$W = \frac{v_0}{2f_0}\sqrt{\frac{2}{\varepsilon_{r+1}}} \tag{3}$$

where v_0 is the velocity of light in air, f_0 is the resonant frequency, and ε_r is the dielectric constant of the substrate. The effective length of the antenna can be calculated by Equations (4 and 5)

$$L_e = L + \Delta L \tag{4}$$

$$L_e = \frac{v_0}{2f_r\sqrt{\varepsilon_{eff}}} \tag{5}$$

$$\text{Again, } L = \frac{\vartheta_0}{2f_r\sqrt{\varepsilon_{eff}}} - 2\Delta L \tag{6}$$

$$\text{where } \varepsilon_{eff} = \frac{\varepsilon_r + 1}{2} + \frac{\varepsilon_r - 1}{2\sqrt{1 + \frac{12h}{w}}} \tag{7}$$

But due to low bandwidth, MSA antennas can't satisfy the desired requirements. A variation of the microstrip antenna is the monopole antenna, which is an ideal candidate to achieve a large bandwidth. The lower frequency of monopole antennas can be calculated by Equations (8–13).

For rectangular-type monopole antennas

$$f_L = \frac{C}{2\left(L_p + W_p\right)\sqrt{\varepsilon_{eff}}} \tag{8}$$

For cylindrical-type monopole antennas

$$L = 0.24\lambda F \tag{9}$$

where $\lambda = \dfrac{L}{L+r}$ putting the value in Equation (9)

$$\lambda = \frac{L+r}{0.24} \tag{10}$$

$$f_l = \frac{C}{\lambda} = \frac{30 \times 0.24}{(L+r)} = \frac{7.2}{(L+r)} \tag{11}$$

If we consider feed length, then the equation becomes

$$f_L = \frac{7.2}{L+r+g} \tag{12}$$

For circular-type monopole antennas:

where L is the length of the rectangular monopole antenna, r is the radius, and g is the probe length. As $L = 2R$ and $r = R/4$; for the quarter-wave monopole antenna, then Equation (12) can be replaced by

$$f_L = \frac{7.2}{2.25R + g} \tag{13}$$

A conventional monopole antenna is difficult to integrate in MMIC. Therefore, a printed planar monopole antenna is used for IC technology. Also, a printed version of a monopole antenna has smaller dimensions and possess similar radiation qualities.

8.2 PERFORMANCE METRICS OF MIMO ANTENNAS

There are four performance parameters to evaluate a MIMO antenna, which are explained in the following.

8.2.1 Envelope Correlation Coefficient

The envelope correlation co-efficient (ECC) indicates how much similarity there is between different received signals. Ideally its value should be zero, but practically it should be as low as possible. In terms of the S-parameter, it can be expressed as follows:

$$ECC = \frac{\left| S_{11}^* S_{12}^* + S_{21}^* S_{22}^* \right|^2}{\left(1 - \left| S_{11} \right|^2 - \left| S_{21} \right|^2\right)\left(1 - \left| S_{22} \right|^2 - \left| S_{21} \right|^2\right)} \tag{14}$$

where S_{11} and S_{22} are the reflection coefficients of the individual antenna, and S_{21} and S_{12} are the transmission coefficients of the individual antenna. This method easily calculates the correlation coefficient of a MIMO antenna.

8.2.2 Diversity Gain

To evaluate the figure-of-merit (FOM) of diversity technique, the term "diversity gain (DG)" is used. It can be mathematically expressed by Equation (15)

$$DG = 10\sqrt{1 - ECC^2} \tag{15}$$

For any MIMO antenna, the relation between ECC and DG are complementary in nature, or in other words it varies inversely to each other. The diversity gain should be high, which means there is a high isolation between antenna elements, which is the desired requirement for a MIMO antenna.

8.2.3 Total Active Refection Coefficient

Another important parameter is total active refection coefficient (TARC), which is used to accurately determine the radiation efficiency and bandwidth of antennas. ECC and DG are not able to determine the accurate radiation efficiency and bandwidth, as they are based on S-parameters. TARC gives more meaningful and complete information about efficiency than the ECC and DG parameters. For a dual-port MIMO system, the TARC can be determined by Equation (16)

$$TARC = \sqrt{\frac{\left(S_{11} + S_{12}\right)^2}{2} + \frac{\left(S_{21} + S_{22}\right)^2}{2}} \tag{16}$$

The value of TARC should be minimal for multiple orientation of a feeding arrangement, which is the indication of efficiency of diversity in MIMO systems.

8.2.4 Channel Capacity Loss

The channel capacity of MIMO systems can be characterized by the channel capacity loss (CCL) parameter for the diversity performance measurement. It helps in defining the loss of transmission bits/s/Hz in a high data-rate transmission. To have good diversity performance for a MIMO antenna, CCL should be smaller than 0.4 bps/Hz.

8.3 CHALLENGES IN MIMO ANTENNA DESIGN

One of the challenges of designing a MIMO antenna is miniaturization. Having multiple antenna elements in a compact size is a great achievement. The spacing between each antenna element should be greater than 0.5 λ to ensure high isolation between them. If antenna elements are spaced less than 0.5 λ apart, then mutual coupling will increase dramatically. There are three types of mutual coupling: near-field coupling, far-field coupling, and surface wave coupling. Near-field coupling is due to the placement of different antenna elements in close proximity. Far-field coupling is if an antenna is placed in the path of far-field zone of another antenna that is about one wavelength distant; then that creates a strong coupling between these two antennas. Surface coupling occurs when a thick substrate is used; then surface waves travel around the discontinuities, like-bends and truncation, and increase the mutual coupling between antennas. Again, due to strong mutual couplings, sometimes destructive interference of fields may occur, which further lead to decrease of overall performance metrics of the MIMO system.

Another challenge is to maintain high isolation among antenna elements so that mutual coupling can be reduced in the MIMO system. That's why designing a multiband MIMO antenna is a cumbersome task. Multiband MIMO antennas are essential today because the same antenna can be used for different frequencies in transmission and reception modes. But multiband MIMO antennas should have high isolation characteristics between various resonating bands.

Again, the size of the MIMO antenna is an important factor, as it affects the mutual coupling, gain, and impedance bandwidth.

There are some applications like ground-penetrating radar and through-wall imaging where a high-gain and high-isolation MIMO system is required. Thus, it is a challenging task to design such multiband MIMO antennas.

8.4 DIFFERENT METHODS TO REDUCE MUTUAL COUPLING

The main problem in MIMO systems is mutual coupling when two or more antennas are in close vicinity, spaced as discussed earlier. Therefore, to design functional MIMO antennas, mutual coupling should be as low as possible,

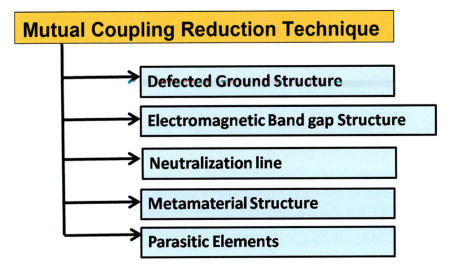

Figure 8.2 Different mutual coupling reduction techniques.

and isolation should be high enough. Mutual coupling mainly arises due to the propagation of surface waves. Suppose a surface wave is traveling in the positive y-direction; then, according to electromagnetic wave theory, it can be expressed by Equation (17)

$$E(y,t) = E_0 e^{jky} \cdot e^{j\omega t} \tag{17}$$

where $k = \omega.\sqrt{\varepsilon\mu}$ is the wave number, E_0 is the amplitude of the surface wave, is the angular frequency, ε is the permittivity of the substrate, μ is permeability, and e^{jwt} is time convention. Here permittivity, permeability, and wave number all are positive, so electromagnetic waves will travel accordingly. The surface waves can be distinguished by creating an artificially periodic structure like meta-material, which has either negative permittivity or negative permeability. When an EM wave passes through, it will disappear instantly. However, for the meta-material there are several methods to reduce the mutual couplings

Researchers have adopted various techniques to reduce mutual coupling, as discussed in the following.

8.4.1 Defected Ground Structure

Many passive and active circuits has adopted a defected ground structure (DGS) in microwave and millimeter wave for band-notched characteristics, concealed surface waves, impedance-matching purposes, and compactness

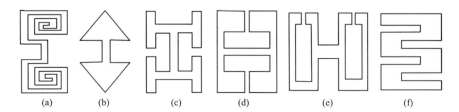

Figure 8.3 Different types of DGS structure: (a) spiral head, (b) arrowhead slot, (c) H-slot, (d) square open loop with a slot in the middle section, (e) open loop dumbbell, (f) inter-digital DGS [3].

of the circuit. DGS gives an extra degree of freedom in the field of microwave and millimeter wave. It is an etched periodic or non-periodic structure in the ground plane of the planar transmission line (microstrip line, coplanar wave guide, etc.). For this etched structure, the current distribution of the structure is disturbed as the characteristics of the transmission line (inductance and capacitance) are changed. Different types of DGS [3], such as spiral head, arrowhead slot, H-slot, square open loop with a slot in the middle section, open loop dumbbell, and inter-digital DGS, are shown in Figure 8.3.

Though DGS has the drawback of a high front-to-back ratio (FBR), there are many more examples for utilizing a defected ground structure in surface wave reduction. Mahmoud et al. used a simple defected ground structure to reduce the mutual coupling between two antennas [4]. Due to the use of the defected ground structure, the coupling isolation was up to −45 dB, though the distance between the two antenna elements was only 1.8 mm. Also, by using a DGS, the correlation coefficient was enhanced by 6.5 dB. The mutual coupling in a microstrip antenna is reduced by implementing a dumbbell-shaped defected ground plane structure in [5].

There is a remarkable reduction of mutual coupling of −34 dB. A novel meander line defected ground plane is added in a two-element MIMO antenna operating at 1.8 GHz [6].

In this work, the author have clearly shown that without DGS, the coupling is −11.1 dB, and with a novel meander line DGS, the mutual coupling is reduced to −15.5 dB at the 1.8 GHz frequency. To increase the isolation between two FIFA antennas for small handheld devices at 2.4 GHz, an extended H-shaped DGS is introduced in [6]. Figure 8.4d shows that an excess of 30 dB mutual coupling reduction is achieved in comparison to an antenna without the defective ground plane structure. Also, the ECC is reduced to 1.3×10^{-4}, which is far below 0.5, the criterion for efficient MIMO characterization. A T-shaped metallic stub is used as a defected ground structure in [7] to increase isolation between antennas in the ultra-wideband range. It can reduce the mutual coupling −20 dB. Also, two open-ended slots are added in the ground plane for the betterment of the impedance matching.

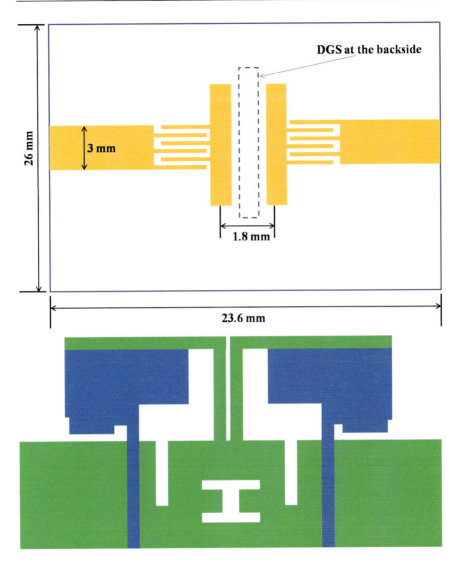

Figure 8.4 MIMO antenna with DGS structure for mitigating mutual coupling factors [4–6]. (a) Simple DGS. (b) Dumbbell-shaped DGS. (c) Meander-line DGS. (d) Extended T-shaped DGS [7].

8.4.2 Electromagnetic Band Gap Structures

An electromagnetic band-gap (EBG) structure is a periodic pattern made on small metal patches or dielectric substrate. The structure mainly prohibits the propagation of EM wave for a frequency band that creates surface waves and thus reduces mutual coupling. There are two different efficient

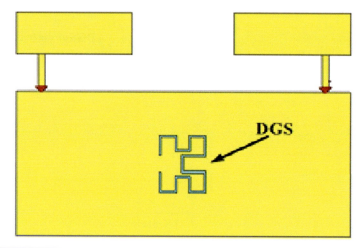

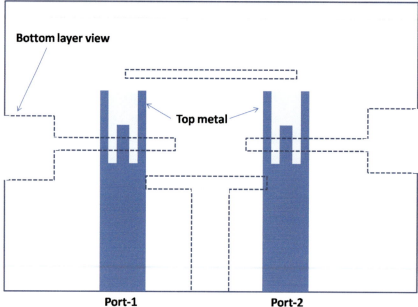

Figure 8.4 Continued

properties of EBG structure; one is phase reflection, and the other is surface wave reduction. The first property is used for antenna gain enhancement, and the second one can be utilized for MIMO antenna decoupling purposes. EBG is a branch of metamaterials that have negative permittivity or permeability that can suppress the surface waves that cause coupling between antenna elements in MIMO systems.

Multiple Input and Multiple Output Antennas 165

Table 8.1 Performance Comparison of Different DGS Embedded MIMO Antennas

Ref.	Freq. Band (GHz)	Substrate (Thickness)	Isolation (dB)	Technique	Diversity Gain (dB)	Size (mm²)	ECC	Application
[4]	5.8	FR4 (1.6 mm)	45	Simple DGS	9.99	2.36 × 2.6	–	–
[5]	3.5	FR4 (0.8 mm)	34	Dumbbell-shaped DGS	4.6	23.4 × 18.2	–	–
[6]	1.8	FR4 (1.6 mm)	15.5	Meander-line DGS	2.78	10 × 14	–	–
[6]	2.4	FR4 (0.8 mm)	30	Extended H-shaped DGS	8.63	60 × 45	1.3×10^{-4}	Small handheld devices
[7]	3.1–11.8	FR4 (0.8 mm)	>20 dB	T-shaped strip and open-ended slot	6	22 × 26	<0.03	UWB

Sometimes an EBG structure needs a large space; then it can exploit the miniaturization of the antenna elements. Therefore, EBG structures should be smaller in dimension with full potentiality.

Different types of EBG structures are added for reduction of mutual coupling, such as a novel type of two-layer electromagnetic band gap structure [8] implemented between two closely spaced planar UWB monopole antennas on a common ground plane. The proposed EBG structures are two closely coupled arrays; one is a linear conducting patch, and other is slit in the ground plane. With this technique, greater than 13 dB isolation is achieved. To get better results, sometimes researchers have implemented two different techniques to increase the isolation between MIMO antennas. Lee et al. [9] introduced both a 1D EBG structure and split ring resonator structure (SRR) simultaneously in one MIMO system. Therefore the mutual coupling was reduced to 53.7 dB, and the back lobes were reduced by 6 dB. Again correlation coefficient is also very minimum of 0.002. Similarly, in 2019, a MIMO antenna was presented with low coupling and high isolation with two hybrid EBG structures and one DGS structure [10]. By incorporating this, the ECC was 0.0006 in simulation and 0.002 in measurement, which made it efficient and usable in a 5G MIMO system. With a combination of three techniques, like EBG, complementary split ring resonator (CSRR), and H-shaped defected ground structure (HDGS), the mutual coupling is reduced by 12 dB in [11]. An unique asymmetric EBG structure with polarization diversity was employed in a compact CPW slot MIMO antenna for coupling reduction purposes. The isolation improved and varied from 26 to 50 dB. The envelope correlation coefficient was less than 0.03 [12].

8.4.3 Neutralization Line

Another way to decouple antenna elements or enhance isolation is to implement a neutralization line (NL). A neutralization line is used to move electromagnetic waves through any slots or slits. The main physics behind it is that, while decoupling, the current cancels the coupling current; then isolation automatically increases. However, this method has a disadvantage in that it increases impedance mismatch and hence incurs extra insertion loss characteristics. Overall antenna efficiency will thus decrease.

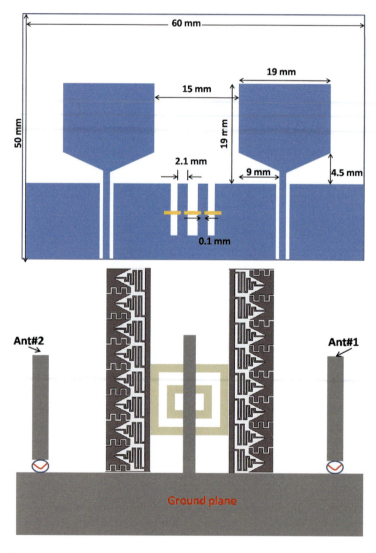

Figure 8.5 MIMO antenna with EBG structure for decoupling. (a) Two-layer EBG. (b) 1D EBG and SRR [8, 9]. (c) Hybrid EBG and DGS [11].

Multiple Input and Multiple Output Antennas

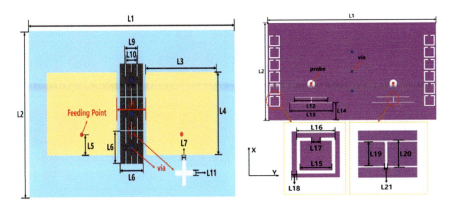

Figure 8.5 Continued

Table 8.2 Performance Comparison of Different EBG-Incorporated MIMO Antennas

Ref	Freq. Band (GHz)	Substrate (Thickness)	Isolation (dB)	Technique	Max Gain/ Efficiency (dB)	Size (mm²)	ECC	Application
[8]	3–6	FR4 (1.5 mm)	>13 dB	Two-layer EBG	90%	60 × 50	–	UWB
[9]	2.46	FR4 (1.2 mm)	53.7	1D EBG+ SRR	82%	60 × 57	0.002	ISM
[10]	4.9	FR4 (1.5 mm)	35	Hybrid EBG+DGS	–	0.88 λ_g × 0.44 λ_g	0.002	5G MIMO
[11]	3.25	FR4 (3 mm)	12	EBG+ CSRR+HDGS	4.9	70 × 40	–	5G MIMO
[12]	3.4–7.3 GHz	FR4 (1.6 mm)	26 to 50	Asymmetric EBG	10	21.6 × 42.2	0.025	sub-6 GHz 5G MIMO

[13] presents a planar UWB MIMO antenna with a compact size of 4 × 4 cm² with a partial slotted ground plane. The structure has in total four numbers of circular-shaped radiators. Out of these, two are in front side and rests are in the back-side. To increase the isolation between them, the front side patches are orthogonally connected to the back side patches. For further reduction of mutual coupling, a stepped neutralization line in the top and bottom layers of the substrate is implemented. The isolation achieved –20 dB of the frequency band from 3.1–11 GHz. In Figure 8.6(b), a dual-antenna system is used for an LTE/smartphone [14] system.

Two neutralization lines are implemented to reduce the mutual coupling by 17.78 dB and to attain a correlation coefficient of less than 0.5. A UWB MIMO antenna achieves –22 dB isolation [15] by introducing a wideband

168 Printed Antennas for Wireless Communication and Healthcare

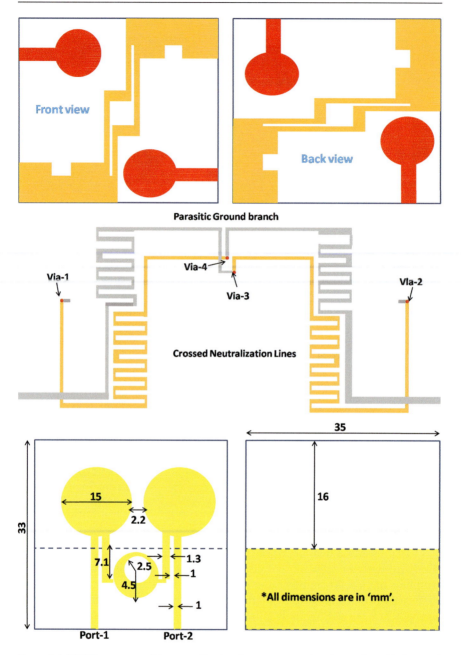

Figure 8.6 MIMO antenna with neutralization line structure for decoupling. (a) Stepped neutralization line. (b) Neutralization line. (c) Two-metal-strip and metal circular neutralization line. (d) U-shaped and inverted U-shaped neutralization line. (e) Systematic neutralization line [13–17].

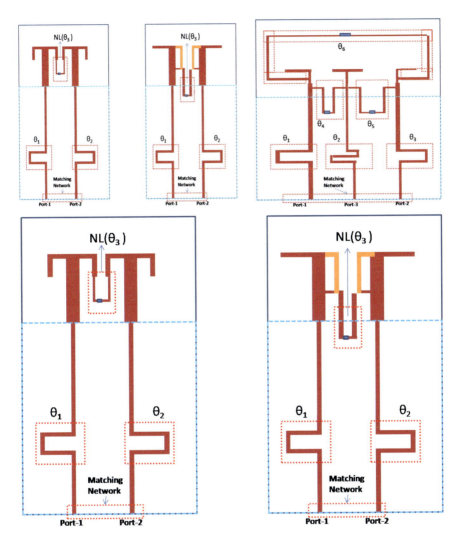

Figure 8.6 Continued

neutralization line between two monopoles. The neutralization line consists of two metal strips and a metal circular disc. The circular disc cancels all possible coupling current paths in different directions in the ground plane. A pair of tri-band MIMO antennas is implemented with high isolation using the NL technique. A U-shaped neutralization line contacts two microstrip lines, [16] which improves the isolation in the high-frequency band 24.3 dB, and an inverted U-shaped NL contacting two radiation patches can decouple in the middle-frequency band 32.4 dB. Recently a systematic approach to

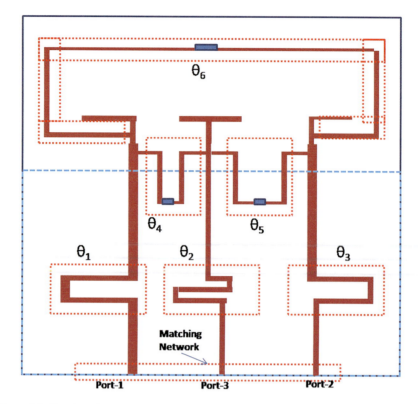

Figure 8.6 Continued

decrease mutual coupling was introduced by using NL [17]. Each NL consists of a metal strip and reactive component attached to its middle portion. With this technique, an isolation of around 25 dB can be obtained with 70 % efficiency.

8.4.4 Metamaterial Structures

From the name itself, it is clear that these are not naturally existing materials or structures. These are artificially engineered structures offering negative permittivity or permeability. Basically, they are represented equivalently by inductance and capacitance, which can work as a band stop filter and can be used as a reduction of unwanted current flow due to surface waves.

There are different types of metamaterial structure, such as meander line (ML), complementary split ring resonator (CSRR), and capacitively loaded loop (CLL). Therefore, it is useful as a band rejection filter for notching out a resonant frequency from a band of frequencies or can be used to increase isolation by reducing the mutual coupling between closely spaced antennas in a MIMO system.

Table 8.3 Performance Comparison of Different MIMO Antennas With Neutralization Lines Implemented

Ref.	Freq. Band (GHz)	Substrate (Thickness)	Isolation (dB)	Technique	Max Gain/Efficiency (dBi)	Size	ECC	Application
[13]	3.1–11	FR4 (1.6 mm)	20	Stepped neutralization line	3.28~4	4×4 cm^2	<0.004	UWB
[14]	0.702–0.968, 1.698–2.216, and 2.264–3	FR4 (0.8 mm)	17.78	Neutralization line	34.20%–43.61% for lower band and 29.63%–61.73% for higher band	135×80 mm^2	<0.5	LTE smartphone
[15]	3.1–5	FR4 (0.8 mm)	22	Two metal strip and metal circular neutralization line	–	35×16 mm^2	<0.1	MIMO
[16]	2.24–2.45 3.3–4 and 5.6- 5.75	FR4 (0.8 mm)	18.2, 32.4 and 24.3	U-shaped and inverted U-shaped neutralization line	–	49×48 mm^2	<0.02	Tri-band MIMO
[17]	2.45	FR4 (0.8 mm)	25	Systematic neutralization line	75.2%–84.0%	95×52 mm^2	0.06~0.03	MIMO array

[18] shows a technique for reducing the mutual coupling of a high-profile monopole antenna by single negative magnetic metamaterial for MIMO application. The MNG strips are placed between closely spaced antennas so to show their effectiveness in reducing mutual coupling and shielding in suppressing displacement current. Another innovative technique to reduce mutual coupling is incorporating a novel metamaterial polarization-rotator (MPR) wall [19]. Actually, the MPR has no effect on the antenna input impedance and radiation pattern. Therefore it can be easily applied between two dielectric resonator antenna array MIMO systems, as shown in Figure 8.7 (b). A capacitively loaded loop metamaterial exhibits a high degree of surface wave attenuation for MIMO antenna application.

In [20], a CLL-MTM superstrate is introduced that achieves 55 dB mutual coupling reduction in a MIMO antenna array system. For microwave imaging application in a UWB MIMO system, [21] used a split ring resonator as a metamaterial to reduce coupling between multiple antennas. By using this material, the performances of the antennas in terms of S-parameter,

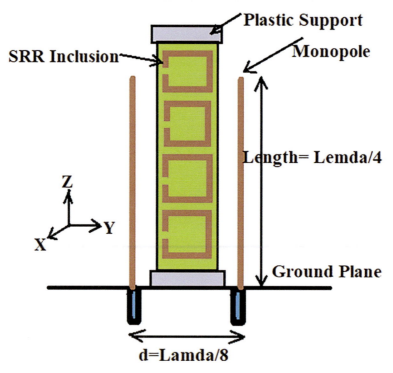

Figure 8.7 MIMO antenna with metamaterial structure for decoupling: (a) single negative magnetic metamaterial (MNG), (b) CSRR metamaterial, (c) CLL-metamaterial superstrate, (d) SRR-metamaterial [18–21].

Multiple Input and Multiple Output Antennas 173

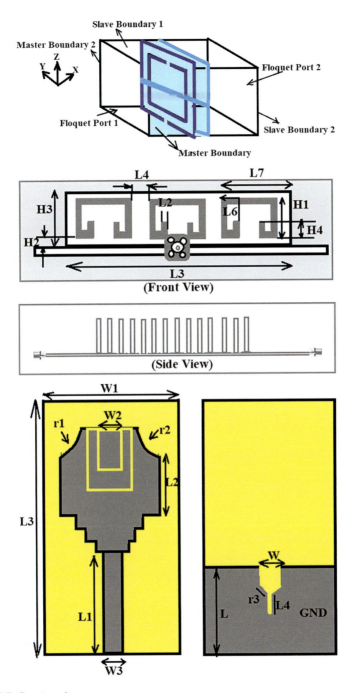

Figure 8.7 Continued

Table 8.4 Performance Comparison of Different Metamaterial MIMO Antennas

Ref.	Freq. Band (GHz)	Substrate (Thickness)	Isolation (dB)	Technique	Max Gain/Efficiency (dBi)	Size (mm²)	ECC	Application
[18]	1.24	Rogers (0.762 mm)	20	Single negative magnetic metamaterial	80~93%	30×70	0.25	MIMO
[19]	57–64	Rogers (RT6010). (10.2 mm)	16	–	88%	1×7	–	MIMO
[20]	3.3	Rogers (2.2 mm)	55	CLL-metamaterial superstrate	8.2 dB and 97%	29.24× 34.7	–	MIMO array
[21]	2–18	FR4 (1.6 mm)	20	SRR-metamaterial	2–8 and 89%	48×35	0.07	MIMO telecommunication
[22]	5.5	FR4 (1 mm)	35	metamaterial absorber	7.53–7.77 dB and 66.64 to 68.03%	9×9	>0.05	MIMO WiMAX

multiplexing efficiency, diversity gain, radiation efficiency, and envelope correlation coefficient improved. Isolation is improved by using a novel flower-shaped metamaterial absorber [22]. It consists of four square rings split at the corners arranged in a symmetrical flower shape. With this system, 35 dB isolation can be achieved with ECC less than 0.05.

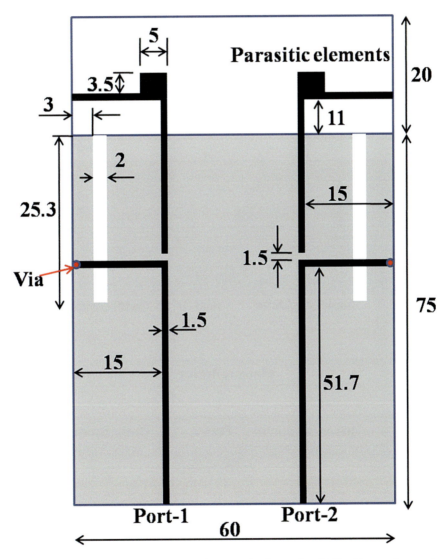

Figure 8.8 MIMO antenna with parasitic elements for decoupling. (a) Double coupling patch parasitic element. (b) Parasitic element. (c) Staircase-shaped radiating element [23–25].

176 Printed Antennas for Wireless Communication and Healthcare

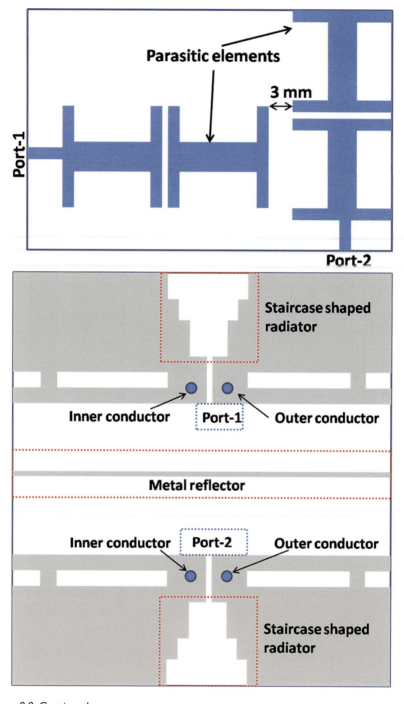

Figure 8.8 Continued

8.4.5 Parasitic Elements

Parasitic elements are placed between multiple MIMO antennas to reduce the mutual coupling between different antennas. The parasitic elements create an opposite current which is induced in the second antenna, and this induced current cancels out the current of the first antenna. Therefore the net mutual coupling current will be zero.

In [23], a double coupling patch is used as a parasitic element in a closely spaced MIMO antenna for mobile terminals. By using this technique, the mutual coupling decreases from −8 to −20 dB. An improvement of 37 dB mutual coupling is observed in [24] by using a parasitic element by employing polarization diversity in an H-shaped patch antenna used for 4G and WiMAX application. In [25], two antenna elements with staircase-type radiators with a bottom-shorted strip are placed at a distance of 0.5 λg. Due to the very short distance, the mutual coupling is very high. To reduce the mutual coupling, a metal conducting strip of 0.4 mm is placed 0.10 to 0.25 λg from the feeding line. This metal strip acts as a reflector and works as a coupling line between two antenna elements.

8.5 TYPES OF MIMO ANTENNAS

The literature reveals that various kinds of MIMO antennas are available depending upon the system requirements. Here we will discuss mainly single/dual-band, multiband, wideband, ultra-wideband, high-gain, and antipodal Vivaldi UWB MIMO antennas. Mutual coupling is the main issue/bottleneck in a MIMO antenna system. Also, there are different techniques to reduce mutual coupling, discussed in the previous section.

Table 8.5 Performance Comparison of Different MIMO Antennas with Parasitic Elements

Ref.	Freq. Band (GHz)	Substrate (Thickness)	Isolation (dB)	Technique	Max Gain/ Efficiency (dBi)	Size (mm²)	ECC	Application
[23]	1.9–2.1	FR4 (0.8 mm)	20	Double coupling patch parasitic element	–	95 × 60	–	MIMO in a mobile terminal
[24]	2.36–5.2 and 7.38	FR4 (1.6 mm)	37	Parasitic element	–	21 × 18	–	4-G and WiMAX application
[25]	3.1–10.6	FR4 (1.6 mm)	20	Staircase-shaped radiating element	5.2 and 90%	25 × 30	–	UWB MIMO

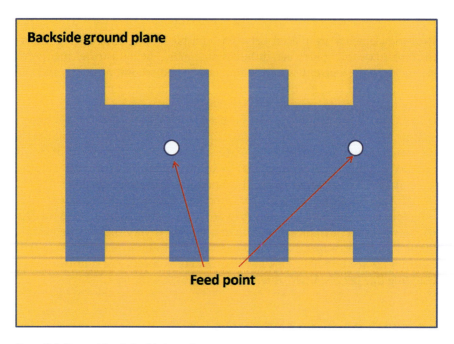

Figure 8.9 Front side of the fabricated antenna.

8.5.1 Single-Band MIMO Antennas

To design a MIMO antenna, microstrip types of antennas are widely used due to their several advantages such as low profile, low cost, ease of fabrication, and ease of manufacture with other monolithic integrated circuits. Figure 8.9 shows an H-shaped microstrip antenna [26] with dumbbell-type DGS at the back side to reduce the mutual coupling.

The antenna is tuned for a single frequency in the 5.8 GHz WLAN band. The dumbbell DGS structure acts as a band stop filter and rectifies the mutual impedance between two antennas. Figure 8.10 (a) shows a planar inverted F antenna [27] used for MIMO techniques at the 28 GHz frequency band. The inverted PIFA antenna has the advantages of high bandwidth compared to the planar antenna and is simple and easy to fabricate. On the other hand Figure 8.10(b) indicates a single-band MIMO antenna for the Ku band for NASA's global positioning system tracking applications [28]. The antenna exhibits a 646 MHz bandwidth with a high gain and isolation characteristics.

8.5.2 Dual/Multiband MIMO Antennas

The digital communication world demands high-speed, small-size, low-profile portable devices. To fulfill these demands, many single-band antennas cannot be designed in limited space. Increasing the number of antennas

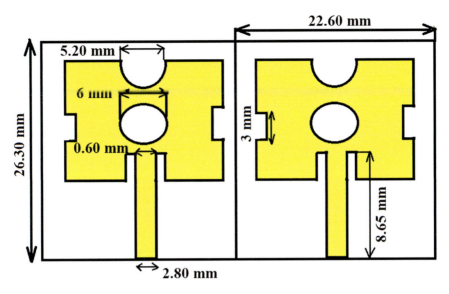

Figure 8.10 Single-band MIMO antennas. (a) Planar inverted F-antenna. (b)MIMO antenna for Ku band [27, 28].

in MIMO systems also degrades the performance due to mutual coupling between them. The solution to this problem is to design a multiband antenna. The authors of [29] designed a four-element MIMO antenna with multiband which covers three LTE bands (LTE700/2300/2500), five WLAN bands (GSM800/950/DCS/PCS/UMTS), and the WLAN 2400 band. An antenna of dimensions 30×26 mm^2 [30] has achieved a dual band of 3.2–3.8 GHz and 5.7–6.2 GHz with higher than −20 dB isolation, as shown in Figure 8.11(b). To improve the isolation, Zicheng et al. designed an antenna with a decoupling structure consisting of a modified array antenna decoupling surface (MADS) [31]. This MADS structure improves the isolation up to −30 dB. To cancel the reactive coupling, Y. Yang et al. designed a MIMO antenna composed of two symmetrical radiating elements [32]. Also, a 3D MIMO antenna structure [33] is fabricated and designed for its multiband performances in four different bands: WLAN, WiMax, 5G cellular, and 5G WiFi. This 3D antenna with compact size has excellent performance for MIMO techniques.

8.5.3 Wide-Band MIMO Antennas

There are several wideband MIMO antennas, which have been designed to enhance the overall bandwidth of the MIMO system. Figure 8.12 shows some of this research work on wideband MIMO antennas. Figure 8.12(a) shows a wideband MIMO antenna where the isolation is improved by etching two

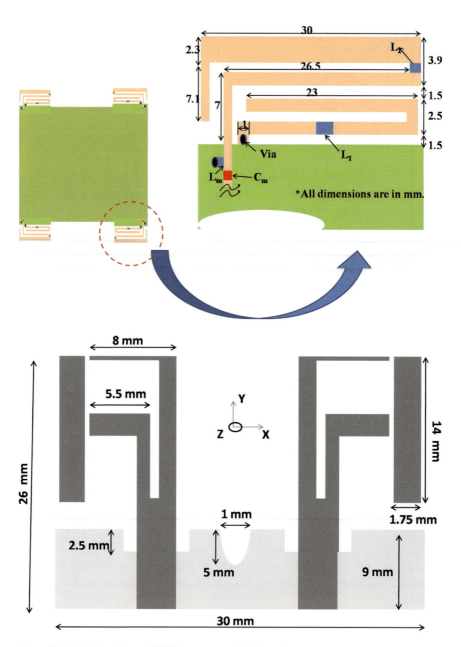

Figure 8.11 Dual/multiband MIMO antennas. (a) Four-element MIMO antenna. (b) Symmetrical structure MIMO. (c) MADS. (d)Multiband MIMO [29–32].

Multiple Input and Multiple Output Antennas 181

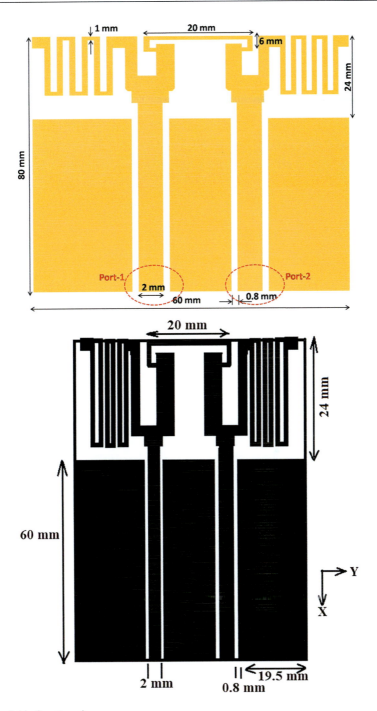

Figure 8.11 Continued

182 Printed Antennas for Wireless Communication and Healthcare

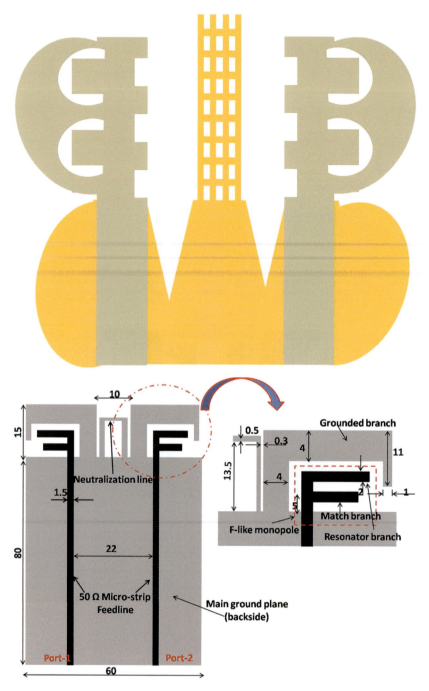

Figure 8.12 Wide-band MIMO antennas. (a) Triangular slot and its S_{11} characteristics. (b) 2.45-GHz operating band antenna. (c) Four-element MIMO antenna [34–36].

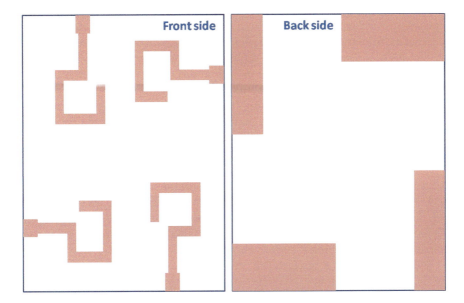

Figure 8.12 Continued

triangular slots at the ground plane and placing a mesh metal strip between the two antennas [34]. The antenna also exhibits a large bandwidth from 2.5 to 12 GHz. Similarly, in [35], a MIMO antenna has a wide bandwidth of 1.09 GHz (1.67 to 2.76 GHz), which covers the applications of GSM1800, GSM1900, UMTS, LTE2300, LTE2500, and 2.45 GHz WLAN band. A wideband meander line 4×4 MIMO antenna [36] was designed with polarization diversity so that it helps to reduce total substrate size without the need of any isolating structure. The antenna displays a bandwidth from 4.76 to 10.6 GHz. Also, a MIMO antenna is designed for 5G applications [37] which cover a bandwidth of 14.5 GHz, from 25.5 to 41 GHz, with a good isolation of 25 dB.

8.5.4 Ultra-Wideband MIMO Antennas

In the last two decades, rapid growth of wireless technology has been witnessed. Now, with the advent of the Internet of Things (IoT), smart homes, smart cities, and smart healthcare systems need wireless access technology with a speed of about 1 Gb/s.

This goal can be reached with the combination of UWB technology and MIMO technology. UWB technology has various merits such as large bandwidth, low cost, multipath fading, and high channel capacity, with an almost 1 Gb/s data rate. Consequently, if an UWB wireless communication system is combined with MIMO technology, then the whole system will be improved with respect to its throughput as well as multipath fading. Hence design of a

UWB-MIMO antenna [38–41] attracts researchers. When designing such a system, researchers mainly focus on the large bandwidth, high isolation characteristics, miniaturization techniques, band notch characteristics, and so on.

For high isolation purposes, different diversity decoupling techniques, slotting, and current neutralization techniques are adopted. For miniaturization purposes, generally, electromagnetic gap structure, fractal geometry, metamaterial structure, meandering line, and defect ground plane structure are implemented. For band notch characteristics, inclusion of split ring structure, electromagnetic band gap structure, cutting slots, resonant structure, fractal geometry structure, and so on are endorsed.

The antenna shown in Figure 8.13 [42] is a small antenna with dimensions of 39 × 39 × 1.6 mm^3. The antenna obtains a wide bandwidth from 2.3–13.75 GHz with triple band notched characteristics by different types of slot, slit, and decoupling structures. Also, the mutual coupling among the antennas is significantly reduced by introducing symmetrical orthogonal separated four-directional staircase structures.

[43] investigated a novel UWB Vivaldi antenna for MIMO application with dual band notched characteristic. The proposed antenna covers the frequency range of 2.9–11.6 GHz with dual band notch characteristics at 5.3–5.8 GHz and a 7.85–8.55 GHz frequency band. On the other hand, a eight-element UWB-MIMO antenna is fabricated for 3G/4G/5G communications [44]. The orthogonal placement of printed monopoles permits polarization diversity and provides high isolation with a broad spectrum from the 2–12 GHz frequency range. A 4×4-element UWB MIMO antenna

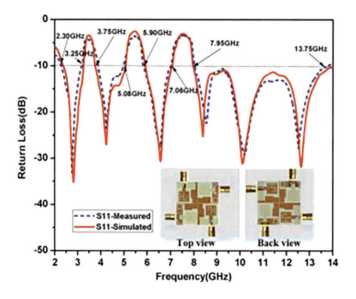

Figure 8.13 Ultra-wideband MIMO antenna [42].

Multiple Input and Multiple Output Antennas   185

creates a dual notch band by incorporating H- and U-slots in a monopole antenna with gap sleeves [45]. This antenna's performance is good, with an envelope correlation coefficient (ECC) less than 0.02, diversity gain (DG) of 10, and peak gain of 5.8 dB, except the notch bands. To reduce coupling of multiple antennas, a metamaterial-based multi-input–multi-output antenna is designed in [46]. The metamaterial is constituted by a linear array of five identical copper unit cells of the SRR type. With this metamaterial structure, the mutual coupling is reduced to 35 dB at 7.8 GHz with an isolation gain of −20 dB compared to antennas without SRR [47].

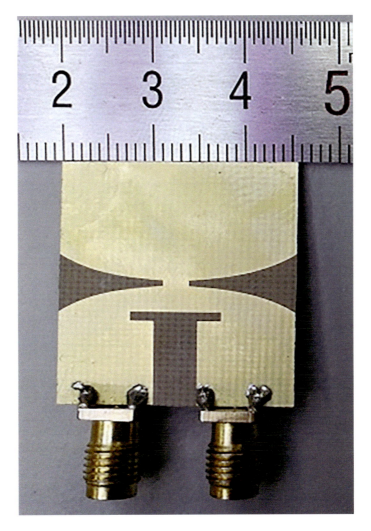

Figure 8.14 Different types of ultra-wideband MIMO antennas. (a) UWB Vivaldi. (b) Eight-element MIMO. (c) Four-element UWB MIMOs (d) Metamaterial-based MIMO. (e) SRR-based MIMO [43–47].

186 Printed Antennas for Wireless Communication and Healthcare

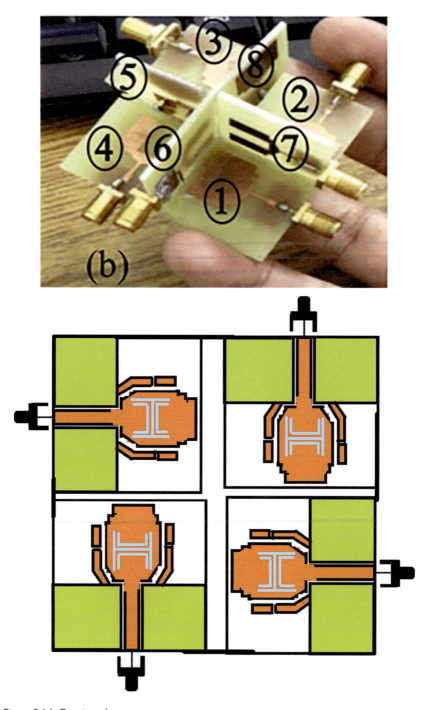

Figure 8.14 Continued

Multiple Input and Multiple Output Antennas 187

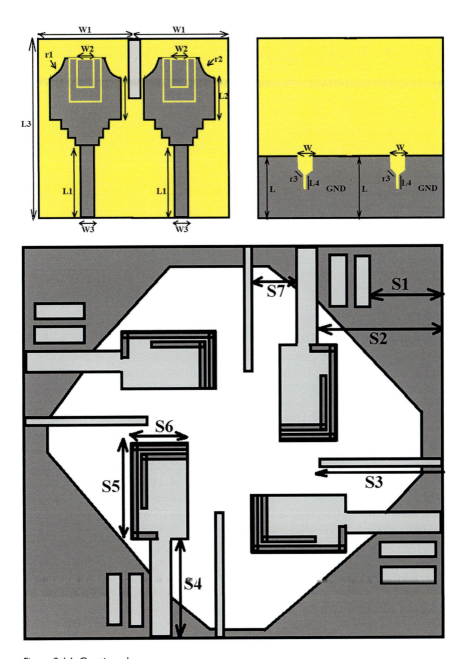

Figure 8.14 Continued

8.6 MIMO ANTENNAS IN BIOMEDICAL USAGE

In every implantable medical device (IMD), a high-resolution image is required for detection and analysis of malignant and benign cells. One promising application of IMD devices is a wireless capsule endoscopy (WCE) system [48, 49]. In the WCE system, the patient swallows a capsule which consists of transmitting antenna, batteries, image sensor, and so on, and outside the body, there are receiving units from which physicians can get the data and images to resolve the particular problem. However, the prime concern in this biotelemetry system is the quality of the captured images and data rate of communication between the transmitting and receiving systems. For acceptable quality of images, a 3.5 Mb/s data rate is required, whereas a 110 Mb/s data rate is mandatory for video frames. The main problem for ultra-wideband technology is that with the increase in frequency range, the propagation path loss also increases.

The multiple-input–multiple-output system offers the high throughout and reliability of IMD devices. Biswas et al. designed a novel type of Hilbert curve [50] fractal MIMO antenna which can serve in wireless capsule endoscopy systems at the 2.45-GHz industrial and scientific band (ISM). The antenna is fabricated on 4-mil-thick liquid crystal polymer substrate (LCP) due to its flexible characteristics so that it can be easily wrapped outside the capsule wall. The antenna size can be made very tiny with the application of fractal geometry. The prototype antenna was measured in liquid performance, showing expected performance at the desired frequency band. MIMO technology is introduced in two orthogonally looped antenna used for an ingestible capsule system [51] to improve its channel capacity. The proposed antenna shows its efficiency for real-time and offline data transmission in a future wireless capsule endoscopy system. An EBG-based implantable four-element MIMO antenna is presented in [52] for implantable medical applications at the 2.4–2.48–GHz ISM band. The electromagnetic band gap structure provides high isolation with an improved gain of −15.18 dBi. Ex vivo measurement in pork tissue showed the antenna's safety and diversity requirements and its potential use in implantable electronics and bio-monitoring applications.

8.7 RECONFIGURABLE MIMO ANTENNAS

MIMO antennas can be used in reconfigurable systems too. Reconfigurability can be done by using a solid-state switch or PIN Diode, RF-MEMS switch, varactor diode, or others.

In [53], a reconfigurable MIMO antenna is used in a WLAN/LTE smartphone application. The antenna satisfactorily covers the hepta band like GSM850/900/GSM1800/1900/UMTS/LTE2300/2500. The antenna presents acceptable performance in all desired respects like efficiency, gain, and

Multiple Input and Multiple Output Antennas 189

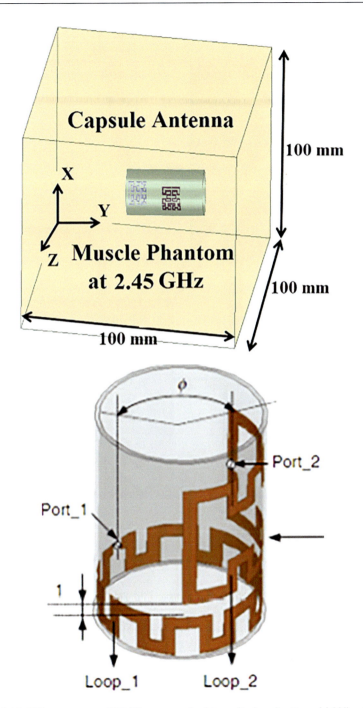

Figure 8.15 Different types of MIMO antennas for biomedical applications. (a) Hilbert curve MIMO. (b) Orthogonally looped MIMO. (c)EBG-based MIMO [50–52].

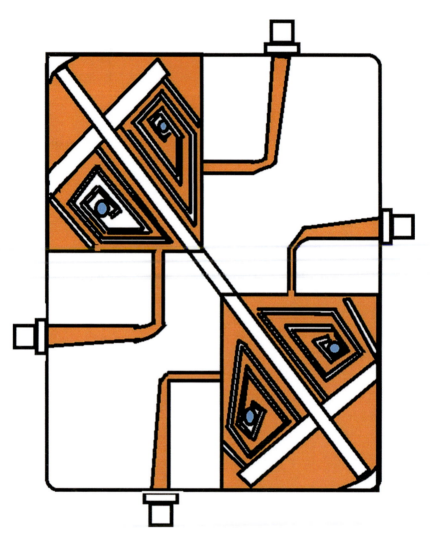

Figure 8.15 Continued

impedance, as well as MIMO parameters like ECC, DG, TARC, ergodic channel capacity, matching, and bandwidth coverage.

One reconfigurable circular patch antenna [54] is presented by exploiting the pattern and polarization diversity to provide the highest gain in a MIMO system in comparison to the standard non-reconfigurable arrays. By using this circular patch antenna, 26% channel capacity is improved. A PIN diode is used here. A novel two-port four-element pattern and frequency reconfigurable MIMO antenna is presented in [55]. The pattern reconfigurability is achieved by a varactor diode and reflecting metal patch.

Multiple Input and Multiple Output Antennas   191

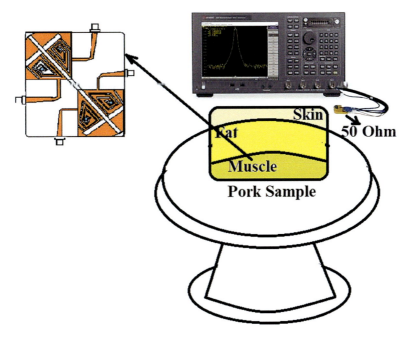

Figure 8.16 Experiment with pork MIMO antenna [52].

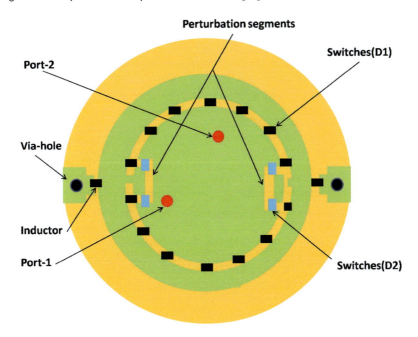

Figure 8.17 Eight element reconfigurable MIMO antenna with ultra-wide band characteristics.

8.8 SUMMARY

This chapter summarizes different types of MIMO antennas and their characteristics, taking into consideration ECC, diversity gain, TARC, and CCL. MIMO antenna application is increasing day by day. Last, the application of MIMO antennas in the biomedical field and reconfigurable MIMO antennas are considered.

REFERENCES

[1] Girish Kumar, and K. P. Ray, *Broadband Microstrip Antennas*, Norwood: Artech House, 2003.

[2] L. Si, H. Jiang, X. Lv, and J. Ding, "Broadband Extremely Close-Spaced 5G MIMO Antenna with Mutual Coupling Reduction Using Metamaterial-Inspired Superstrate," *Optic Express*, Vol. 27, no. 3, pp. 3472–3482, 2019.

[3] L. H. Wang, Y. C. Guo, X. W. Shi, and X. Q. Chen "An Overview on Defected Ground Structure," progress in electromagnetic research B, Vol. 7 pp. 173–179, 2008.

[4] M. A. Abdalla, and A. A. Ibrahim, "Compact and Closely Spaced Metamaterial MIMO Antenna With High Isolation for Wireless Applications," *IEEE Antennas and Wireless Propagation Letters*, Vol. 12, pp. 1452–1455, 2013, doi: 10.1109/LAWP.2013.2288338.

[5] A. Kumar, and S. Devane, "DGS Based Mutual Coupling Reduction between Microstrip Patch Antenna Arrays for WIMAX Applications," *2016 2nd International Conference on Next Generation Computing Technologies* (NGCT), pp. 78–83, 2016, doi: 10.1109/NGCT.2016.7877393.

[6] A. H. Abdelgwad, and M. Ali, "Mutual Coupling Reduction of a Two-element MIMO Antenna System Using Defected Ground Structure," *2020 IEEE International Symposium on Antennas and Propagation and North American Radio Science Meeting*, pp. 1909–1910, 2020, doi: 10.1109/IEEECONF35879.2020.9330404.

[7] C. R. Jetti, and V. R. Nandanavanam, "Trident-Shape Strip Loaded Dual Band-Notched UWB MIMO Antenna for Portable Device Applications," *AEU-International Journal of Electronics and Communications*, Vol. 83, pp. 11–21, 2018.

[8] Q. Li, A. P. Feresidis, M. Mavridou, and P. S. Hall, "Miniaturized Double-Layer EBG Structures for Broadband Mutual Coupling Reduction Between UWB Monopoles," *IEEE Transactions on Antennas and Propagation*, Vol. 63, no. 3, pp. 1168–1171, 2015, doi: 10.1109/TAP.2014.2387871.

[9] J. Y. Lee, S. H. Kim, and J. H. Jang, "Reduction of Mutual Coupling in Planar Multiple Antenna by Using 1-D EBG and SRR Structures," *IEEE Transactions on Antennas and Propagation*, Vol. 63, no. 9, pp. 4194–4198, 2015.

[10] Y. Liu, X. Yang, Y. Jia, and Y. J. Guo, "A Low Correlation and Mutual Coupling MIMO Antenna," *IEEE Access*, Vol. 7, pp. 127384–127392, 2019, doi: 10.1109/ACCESS.2019.2939270.

[11] Z. Yang, J. Xiao, and Q. Ye, "Enhancing MIMO Antenna Isolation Characteristic by Manipulating the Propagation of Surface Wave," *IEEE Access*, Vol. 8, pp. 115572–115581, 2020, doi: 10.1109/ACCESS.2020.3004467.

[12] A. I. Afifi, A. S. Abd El-Hameed, A. Allam, S. M. Ahmed, and A. B. Abdel-Rahman, "Dual Port MIMO Antenna with Low Mutual Coupling Based on Asymmetric EBG Decoupling Structure," *2021 15th European Conference on Antennas and Propagation* (EuCAP), pp. 1–5, 2021, doi: 10.23919/EuCAP51087.2021.9411149.

[13] W. A. E. Ali, and A. A. Ibrahim, "A Compact Double-Sided MIMO Antenna with an Improved Isolation for UWB Applications," *AEU-International Journal of Electronics Communication*, Vol. 82, pp. 7–13, 2017.

[14] S. Wang, and Z. Du, "Decoupled Dual-Antenna System Using Crossed Neutralization Lines for LTE/WWAN Smartphone Applications," *IEEE Antennas and Wireless Propagation Letters*, Vol. 14, pp. 523–526, 2015, doi: 10.1109/LAWP.2014.2371020.

[15] S. Zhang, and G. F. Pedersen, "Mutual Coupling Reduction for UWB MIMO Antennas With a Wideband Neutralization Line," *IEEE Antennas and Wireless Propagation Letters*, Vol. 15, pp. 166–169, 2016, doi: 10.1109/LAWP.2015.2435992.

[16] R. Liu, X. An, H. Zheng, M. Wang, Z. Gao, and E. Li, "Neutralization Line Decoupling Tri-Band Multiple-Input Multiple-Output Antenna Design," *IEEE Access*, Vol. 8, pp. 27018–27026, 2020, doi: 10.1109/ACCESS.2020.2971038.

[17] M. Li, L. Jiang, and K. L. Yeung, "A General and Systematic Method to Design Neutralization Lines for Isolation Enhancement in MIMO Antenna Arrays," *IEEE Transactions on Vehicular Technology*, Vol. 69, no. 6, pp. 6242–6253, 2020, doi: 10.1109/TVT.2020.2984044.

[18] M. M. Bait-Suwailam, M. S. Boybay, and O. M. Ramahi, "Electromagnetic Coupling Reduction in High-Profile Monopole Antennas Using Single-Negative Magnetic Metamaterials for MIMO Applications," *IEEE Transactions on Antennas and Propagation*, Vol. 58, no. 9, pp. 2894–2902, 2010, doi: 10.1109/TAP.2010.2052560.

[19] M. Farahani, J. Pourahmadazar, M. Akbari, M. Nedil, A. R. Sebak, and T. A. Denidni, "Mutual Coupling Reduction in Millimeter-Wave MIMO Antenna Array Using a Metamaterial Polarization-Rotator Wall," *IEEE Antennas and Wireless Propagation Letters*, Vol. 16, pp. 2324–2327, 2017, doi: 10.1109/LAWP.2017.2717404.

[20] A. Jafargholi, A. Jafargholi, and J. H. Choi, "Mutual Coupling Reduction in an Array of Patch Antennas Using CLL Metamaterial Superstrate for MIMO Applications," *IEEE Transactions on Antennas and Propagation*, vol. 67, no. 1, pp. 179–189, 2019, doi: 10.1109/TAP.2018.2874747.

[21] H. Sakli, C. Abdelhamid, C. Essid, and N. Sakli, "Metamaterial-Based Antenna Performance Enhancement for MIMO System Applications," *IEEE Access*, Vol. 9, pp. 38546–38556, 2021, doi: 10.1109/ACCESS.2021.3063630.

[22] P. Garg, and P. Jain, "Isolation Improvement of MIMO Antenna Using a Novel Flower Shaped Metamaterial Absorber at 5.5 GHz WiMAX Band," *IEEE Transactions on Circuits and Systems II: Express Briefs*, Vol. 67, no. 4, pp. 675–679, 2020, doi: 10.1109/TCSII.2019.2925148.

[23] Z. Li, Z. Du, M. Takahashi, S. Saito, and K. Ito, "Reducing Mutual Coupling of MIMO Antennas with Parasitic Elements for Mobile Terminals", *IEEE Transactions Antennas Propagation*, Vol. 60, no. 2, pp. 473–481, 2012.

[24] R. Ghoname, and A. Zekry, "Mutual Coupling Reduction of MIMO Antennas using Parasitic Elements for Wireless Communications," *International Journal of Computer Application*, Vol. 62, No. 19, 2013.

[25] T. K. Roshna, U. Deepak, V. R. Sajitha, K. Vasudevan, and P. Mohanan, "A compact UWB MIMO antenna with reflector to enhance isolation," *IEEE Transactions Antennas Propagation*, Vol. 63, no. 4, pp. 1873–1877, 2015.

[26] N. Pratima, and N. Anil, "A Compact H Shaped MIMO Antenna with an Improved Isolation for WLAN Applications," in *Advanced Computing and Communication Technologies*, Singapore: Springer, pp. 493–501, 2016, doi: 10.1007/978-981-10-1023-1_49.

[27] M. Haraz, M. Ashraf, and S. Alshebeili, "Single-Band PIFA MIMO Antenna System Design for Future 5G Wireless Communication Applications," *2015 IEEE 11th International Conference on Wireless and Mobile Computing, Networking and Communications* (WiMob), pp. 608–612, 2015, doi: 10.1109/WiMOB.2015.7348018.

[28] H. S. Gill, S. Singh, M. Singh, and G. Kaur, "Design of Single-Band MIMO Antenna For KU-Band Applications," *2019 International Conference on Electrical, Electronics and Computer Engineering* (UPCON), pp. 1–5, 2019, doi: 10.1109/UPCON47278.2019.8980185.

[29] A. Ahmad, and F. A. Tahir, "Multiband MIMO Antenna on Variable-Sized Tablet PCs," *2017 IEEE Asia Pacific Microwave Conference* (APMC), pp. 612–615, 2017, doi: 10.1109/APMC.2017.8251520.

[30] Z. Niu, H. Zhang, Q. Chen, and T. Zhong, "Isolation Enhancement in Closely Coupled Dual-Band MIMO Patch Antennas," *IEEE Antennas and Wireless Propagation Letters*, Vol. 18, no. 8, pp. 1686–1690, 2019, doi: 10.1109/LAWP.2019.2928230.

[31] P. C. Nirmal, A. Nandgaonkar, S. L. Nalbalwar, and R. Kumar Gupta, "A Compact Dual Band MIMO Antenna with Improved Isolation for Wi-MAX and WLAN Applications," *Progress In Electromagnetics Research M*, Vol. 68, pp. 69–77, 2018, doi: 10.2528/PIERM18033104

[32] Y. Yang, Q. Chu, and C. Mao, "Multiband MIMO Antenna for GSM, DCS, and LTE Indoor Applications," *IEEE Antennas and Wireless Propagation Letters*, Vol. 15, pp. 1573–1576, 2016, doi: 10.1109/LAWP.2016.2517188.

[33] R. Saleem, M. Bilal, H. T. Chattha, S. Ur Rehman, A. Mushtaq, and M. F. Shafique, "An FSS Based Multiband MIMO System Incorporating 3D Antennas for WLAN/WiMAX/5G Cellular and 5G Wi-Fi Applications," *IEEE Access*, Vol. 7, pp. 144732–144740, 2019, doi: 10.1109/ACCESS.2019.2945810.

[34] P. C. Nirmal, N. Anil, N. Sanjay, and R. K. Gupta, "Compact Wideband MIMO Antenna for 4G WI-MAX, WLAN and UWB Applications," *AEU—International Journal of Electronics and Communications*, Vol. 99, pp. 284–292, 2019, doi: 10.1016/j.aeue.2018.12.008.

[35] Y. Wang, and Z. Du, "A Wideband Printed Dual-Antenna System With a Novel Neutralization Line for Mobile Terminals," *IEEE Antennas and Wireless Propagation Letters*, Vol. 12, pp. 1428–1431, 2013, doi: 10.1109/LAWP.2013.2287199.

[36] L. Malviya, K. Gehlod, and A. Shakya, "Wide-Band Meander Line MIMO Antenna For Wireless Applications," *2018 International Conference on Advances in Computing, Communications and Informatics* (ICACCI), pp. 1663–1667, 2018, doi: 10.1109/ICACCI.2018.8554719.

[37] A. Ejaz, S. Mehak, W. Anwer, Y. Amin, J. Loo, and H. Tenhunen, "Investigating a 28 GHz Wide-Band Antenna and its MIMO Configuration," *2019 2nd International Conference on Communication, Computing and Digital systems* (C-CODE), pp. 7–10, 2019, doi: 10.1109/C-CODE.2019.8680992.

[38] Yi Zhao, Fu-Shun Zhang, Li-Xin Cao, and Deng-Hui Li, "A Compact Dual Band Notch MIMO Diversity Antenna for UWB Wireless Application," *Progress in Electromagnetics Research C*, Vol. 89, pp. 161–169, 2019 doi: 10.2528/PIERC18111902.

[39] N. K. Kiem, H. N. Bao Phuong, and D. N. Chien, "Design of Compact 4 × 4 UWB-MIMO Antenna with WLAN Band Rejection," *Hindawi Publishing Corporation International Journal of Antennas and Propagation*, Vol. 2014, Article ID: 539094, p. 11, doi: 10.1155/2014/539094.

[40] A. Mchbal, N. Amar Touhami, H. Elftouh, and A. Dkiouak, "Mutual Coupling Reduction Using a Protruded Ground Branch Structure in a Compact UWB Owl-Shaped MIMO Antenna," *Hindawi International Journal of Antennas and Propagation*, Vol. 2018, Article ID: 4598527, p. 10, doi: 10.1155/2018/4598527.

[41] H. T. Chattha, F. Latif, F. A. Tahir, M. U. Khan, and X. Yang, "Small-Sized UWB MIMO Antenna with Band Rejection Capability," *IEEE Access*, Vol. 7, pp. 121816–121824, 2019, doi: 10.1109/ACCESS.2019.2937322.

[42] Z. Tang, X. Wu, J. Zhan, S. Hu, Z. Xi, and Y. Liu, "Compact UWB-MIMO Antenna with High Isolation and Triple Band-Notched Characteristics," *IEEE Access*, Vol. 7, pp. 19856–19865, 2019, doi: 10.1109/ACCESS.2019.2897170.

[43] Z. Li, C. Yin, and X. Zhu, "Compact UWB MIMO Vivaldi Antenna with Dual Band-Notched Characteristics," *IEEE Access*, Vol. 7, pp. 38696–38701, 2019, doi: 10.1109/ACCESS.2019.2906338.

[44] M. S. Khan, A. Iftikhar, R. M. Shubair, A. Capobianco, B. D. Braaten, and D. E. Anagnostou, "Eight-Element Compact UWB-MIMO/Diversity Antenna with WLAN Band Rejection for 3G/4G/5G Communications," *IEEE Open Journal of Antennas and Propagation*, Vol. 1, pp. 196–206, 2020, doi: 10.1109/OJAP.2020.2991522.

[45] V. S. D. Rekha, P. Pardhasaradhi, B. T. P. Madhav, and Y. U. Devi, "Dual Band Notched Orthogonal 4-Element MIMO Antenna with Isolation for UWB Applications," *IEEE Access*, Vol. 8, pp. 145871–145880, 2020, doi: 10.1109/ACCESS.2020.3015020.

[46] H. Sakli, C. Abdelhamid, C. Essid, and N. Sakli, "Metamaterial-Based Antenna Performance Enhancement for MIMO System Applications," *IEEE Access*, Vol. 9, pp. 38546–38556, 2021, doi: 10.1109/ACCESS.2021.3063630.

[47] Z. Chen, W. Zhou, and J. Hong, "A Miniaturized MIMO Antenna with Triple Band-Notched Characteristics for UWB Applications," *IEEE Access*, Vol. 9, pp. 63646–63655, 2021, doi: 10.1109/ACCESS.2021.3074511.

[48] Biswas B, Karmakar A, Chandra V. "Miniaturized Wideband Ingestible Antenna for Wireless Capsule Endoscopy," *IET Microwaves, Antennas and Propagation*, Vol. 14, no. 4, pp. 293–301, 2019.

[49] B. Biswas, A. Karmakar, and V. Chandra, "Fractal Inspired Miniaturized Wideband Ingestible Antenna for Wireless Capsule Endoscopy," *International Journal of Electronics Communication* (AEÜ), Vol. 120, 2020.

[50] B. Biswas, A. Karmakar, and V. Chandra, "Hilbert Curve Inspired Miniaturized MIMO Antenna for Wireless Capsule Endoscopy," *International Journal of Electronics Communication* (AEÜ), Vol. 137, 2021.

[51] L. J. Xu, B. Li, M. Zhang, and B. Yaming, "Conformal MIMO Loop Antenna for Ingestible Capsule Applications," *Electronics Letters*, Vol. 53, no. 23, pp. 1506–1508, 2017.

[52] Y. Fan, J. Huang, T. Chang, and X. Liu, "A Miniaturized Four-Element MIMO Antenna With EBG for Implantable Medical Devices," *IEEE Journal of Electromagnetics, RF and Microwaves in Medicine and Biology*, Vol. 2, no. 4, pp. 226–233, 2018, doi: 10.1109/JERM.2018.2871458.

[53] Z. Xu, Y. Sun, Q. Zhou, Y. Ban, Y. Li, and S. S. Ang, "Reconfigurable MIMO Antenna for Integrated-Metal-Rimmed Smartphone Applications," *IEEE Access*, Vol. 5, pp. 21223–21228, 2017, doi: 10.1109/ACCESS.2017.2757949.

[54] D. Piazza, P. Mookiah, M. D'Amico, and K. R. Dandekar, "Pattern and Polarization Reconfigurable Circular Patch for MIMO Systems," *2009 3rd European Conference on Antennas and Propagation*, pp. 1047–1051, 2009.

[55] A. H. Abdelgwad, and M. Ali, "Frequency/Pattern Reconfigurable Printed Monopole MIMO Antenna for Handheld Devices," *2021 15th European Conference on Antennas and Propagation* (EuCAP), pp. 1–4, 2021, doi: 10.23919/EuCAP51087.2021.9411487.

Problems

1. What is MIMO technology?
2. What are the different performance metrics of MIMO systems?
3. How can you minimize the mutual coupling between antenna elements of a MIMO system?
4. What are ECC, TARC, and DG? How they are related?
5. What is the channel information matrix (H) of a MIMO system?
6. How are the limitations of the Shannon-Hartley theorem overcome using MIMO systems?
7. What kind of diversity technique is implemented in MIMO systems? Explain with a diagram.
8. What are the challenges in designing efficient MIMO antennas?
9. What is a reconfigurable antenna? In what respect can an antenna be reconfigured? Can a MIMO system be realized by implementing reconfigurable antennas?
10. Can a MIMO system be utilized for biomedical applications?
11. What are the salient features of a MIMO system?
12. What is meta-material? How can it be implemented for MIMO antennas?
13. What is massive MIMO? Is there any practical use for such a system?

14. What are DGS and EBG structure? Explain with a proper diagram. Can they be used for a MIMO antenna?
15. How can S-parameters be utilized to evaluate the figure-of-merit of a MIMO system?
16. What are the different types of antennas generally used for MIMO systems?

Chapter 9

Antennas for Microwave Imaging

9.1 INTRODUCTION

Microwave imaging is a special type of technology that exploits the radiation of frequency for detecting and locating hidden, concealed, or embedded objects using electromagnetic waves (EMWs) in the microwave region (300 MHz to 300 GHz). Based on the frequency, the radiation can penetrate through many optically opaque media, like wall, cloth, soil, fog, wood, concrete, and plastic. Hardware and software components are required for microwave imaging purposes. The transmitting antenna transmits EM waves to the sample under test (different parts of the human body for medical treatment/concealed weapons). The hardware component collects data from the antenna. If the sample is made up of homogeneous material, and then no EM wave will be reflected. However, if the sample has any anomalies which may have different electrical or magnetic properties, then a portion of the EM wave will be reflected back, and it will be stronger if the change in property between the anomaly and surrounding environment becomes prominent. The reflected data can be collected by the same transmitting antenna by mono-static method or by any other receiver antenna by the bi-static method. The collected raw data is fed to different software algorithms for processing purposes. Based upon the software algorithm, microwave imaging is categorized into two parts, quantitative and qualitative. The qualitative method calculates a qualitative profile of the hidden objects, whereas the quantitative method exactly calculates the size, location, and orientation of the hidden objects. The synthetic aperture radar (SAR) and ground penetrating radar (GPR) algorithms are two of the most popular algorithms for qualitative imaging. Table 9.1 gives the details of the two methods.

9.1.2 Why Microwave?

Medical imaging is widely used for visualizing the internal human body. It can be used from all phases of medical treatment from bone fracture to cancer treatment; its application is widespread. There are different techniques

DOI: 10.1201/9781003389859-9

199

200 Printed Antennas for Wireless Communication and Healthcare

Table 9.1 Comparison of Algorithms Between Qualitative and Quantitative

Types	Method	Algorithm	Frequency	Positive Aspects	Negative Aspects
Qualitative	Radar scattering technique	1. Synthetic aperture radar 2. Ground-penetrating radar	UWB frequency range 3.1–10.6 GHz	It is easy to compute and requires less time and cost	It is unable to provide exactly the location and size of the object
Quantitative	Inverse electromagnetic scattering methods used iteratively	1. Gauss-Newton method 2. Log magnitude and phase reconstruction method	Single-frequency, multiple-frequency etc.	Exactly provides the object's shape, size, and location	When using direct inverse method, the process is costly, and the iterative method is time consuming

used for microwave imaging such as X-ray, magnetic resonance imaging (MRI), ultrasound graph (USG), computed tomography (CT scan), and position emission tomography (PET), which are the most popular. These different methods have provided a variety of resolutions, implementation costs, complexity, and ionizing radiation. Among them, MRI provides good resolution at the expense of higher cost and patient discomfort with the procedure. Similarly, a CT scan has high resolution but provides high radiation exposure and is less informative. X-rays can ionize the tissue and increase health risk. PET can yield good information for soft tissue but suffers from poor resolution. So the major problem with the existing systems are safety, cost, and accuracy. The alternative to the aforementioned technologies is microwave frequencies of non-ionizing electromagnetic waves for sensing and imaging for medical diagnosis at low cost and low health risk with accurate measurement.

9.1.3 Challenges to Designing Microwave Imaging Systems

Over many years, designers have implemented microwave imaging techniques for medical purposes for early detection and diagnosis of cancer. However, there are many challenges to face during the design of microwaves for human body imaging and the conversion of data to high-resolution images.

1. Most of the antennas, designed for parcatical usages are assumed to be in free-space while evaluating its performance metrics. However,

antennas used for medical imaging have a conductive environment which has different conductivity and permittivity depending upon the tissue properties and temperature used. The human body consists of different body tissues which have different electrical properties. The electrical properties of stomach tissue are totally different from the colon or muscle tissue. So it is important to avoid an air–body interface mismatch.

2. The bandwidth of the antenna should be high enough to avoid the detuning effect. Narrow-band antennas can't be used for this purpose.

3. Human tissue is heterogeneous in nature and also dispersive, which causes multiple scatters to appear as clutter. A strong imaging algorithm can only compensate for clutter. Also, signals are highly attenuated due to lossy human tissue. It also affects the depth of penetration.

4. An antenna array offers higher performance in microwave imaging, though it requires a large size. Various techniques are adopted to miniaturize the antenna array. But with high frequency, the depth of penetration of EM waves will be decreased.

5. By increasing the number of antenna elements in a array, the performance will automatically increase so that the system can create an image which will closely resemble the original one. However, simultaneously, it will increase the total area and mutual coupling between array elements. Again, due to the increase of mutual coupling, the performance of the system will automatically be decreased.

6. A suitable choice of frequency is also a challenging task because it is a trade-off between penetration depth and resolution. It is a well-known fact that as the frequency increases, the penetration depth decreases and tissue attenuation increases. However, with increased frequency, the resolution of the images increases. So, depending upon the application, the frequency should be chosen. For detection of concealed weapons, where high penetration depth is required, low frequency is required, and for detection of breast cancer, where high resolution is essential, a high frequency should be chosen.

7. Microwave imaging is based on the significant difference of the electrical properties between malignant and normal tissue. However, this imaging technique may not work faithfully while very slight difference in electrical property is observed between the benign and malignant tissues.

8. For sensing applications like heartbeat detection or respiration artifact, the background clutter creates noise. It is a very challenging task to separate the desired sound from the background clutter. Hence, a robust, efficient signal processing algorithm is required.

9. The material of the antenna should be bio-comfortable in nature so that it will not hamper the human body.

10. Data acquisition and processing techniques should have high resolution to produce the correct size and location of the targeted image data so that false detection can be avoided. It is expected that much time will be needed to process the data and generate images.
11. The imaging algorithm should be efficient enough to show the exact location and size of the targeted data. A 2D-based imaging algorithm may not work for a 3D biological object; it demands a 3D-based imaging algorithm. Each algorithm has its own limitations. Properly selecting the algorithm for a particular application is important work, so in-depth studies are required to develop efficient and accurate computational algorithms.
12. The surrounding interference, like mobile phones, computers, fluorescent lights, magnetic fields, static discharge, and electrical power supplies, can interfere with microwave imaging systems, so microwave imaging systems should be used in a particular room free from these disturbances.
13. Besides the technical challenges, there are some non-technical challenges. Though microwave imaging has the advantages of low risk, time resolution, and cost effectiveness, it lacks special resolution in comparison to MRI, CT scan, and so on.

9.1.4 Application of Microwave Imaging

There are variety of applications for microwave imaging.

1. Medical imaging

A decade ago, medical imaging saw a revolutionary metamorphosis by producing pictures inside the human body. There are different methods used for imaging purposes, such as X-ray, computer tomography scan, magnetic resonance imaging, ultrasound imaging, and position emission tomography. For early detection of breast tumors or cancer, microwave medical imaging is an important and non-invasive technique. Malignant tissue in the human body contains more water and blood than normal tissue. Therefore, EM wave scatters from that tissue can be identified by the imaging technique.

2. Concealed weapon detection at security checkpoints

In high-security areas such as airports, government buildings, and courthouses, it is essential to monitor safety. Sometimes it is not possible to detect concealed weapons carried by individuals by metal detectors. Microwave frequency has a inherent characteristic of penetrating any objects like cloth, metal, or the body. By properly using this unique property, microwave imaging techniques can efficiently detect concealed weapons.

3. Structural health monitoring

Aging materials/buildings are now a serious concern in the world. Infrastructure becomes corroded due to rain or sunshine over time. Its longevity is thus shortened. Recently microwave imaging has brought tremendous improvement in structural health monitoring. Low-frequency microwave signals (less than 10 GHz) can easily penetrate concrete and detect if there is any rust or corrosion. Microwave imaging can also detect if there are any cracks or air voids in cement.

4. Through-wall imaging

The ability to see through obstacles, such as walls, windows, and other optically opaque things, using microwave signals has become a major area of interest in a variety of applications for both military and commercial purposes. It is applicable for law-enforcement officials, counter-terrorism agents, search-and-rescue workers, and those who encounter situations where they need to detect, locate, and identify occupants of a building. Through-wall imaging can be a powerful tool for that use.

5. Weather radar

Radio detection and ranging (radar) works in the microwave frequency range for detection and location of objects in the air. Radar transmits a focused-pulse electromagnetic wave; then part of this wave is reflected after hitting the target object. From this reflection of the energy, it is capable of detecting objects like rain, snow, and storms. Thus, microwave imaging systems can help by predicting the severity of storms.

9.2 TYPES OF MICROWAVE IMAGING

Microwave imaging can be classified into three categories: passive, hybrid, and active. For suitable microwave imaging, especially for breast cancer detection, these three techniques are used.

For the case of passive microwave imaging in breast cancer detection, the temperature difference between normal and tumor breast tissue can be observed. Cancerous or tumor tissue generally has absorbed more energy and is more heated than normal breast tissue. A hybrid technique needs an ultrasound transducer. Malignant tissue has higher conductivity, so it absorbs more energy and expands. The expansion creates a pressure wave, and this wave can be detected by the ultrasound transducer. In active microwave imaging, antennas are essential. The antenna is used for transmitting a low-power microwave signal pulse to penetrate human tissue and receive

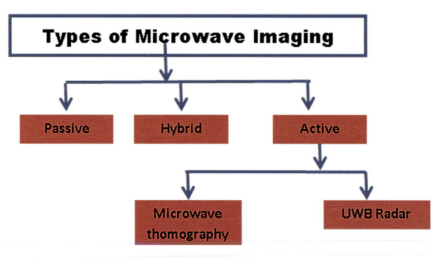

Figure 9.1 Different types of microwave imaging.

the back-scattered signal response. Two active process are employed for microwave imaging, microwave tomography and UWB radar-based imaging. In the microwave tomography technique, the antenna collects data from the surrounding malignant tissues. Then a forward and reverse electromagnetic problem can be solved to detect and locate the cancerous cell. In this approach antennas covering a low-frequency range (1–6 GHz) are needed, as they are efficient for penetrating tissue easily. In the UWB radar technique, an antenna or a set of antennas (an antenna array) is applied for transmitting and receiving short pulses from different locations. The exact size and location can be determined by the time delay between the transmitted pulse and scattered signals and also from the amplitude of the scattered signals. There are different imaging algorithms such as space time beam forming, tissue-sensing adaptive radar, delay, and sum imaging algorithms [1–3]. In this approach, antennas for high-frequency coverage are applicable because high-resolution image construction is required here. In comparison to both active microwave imaging techniques, the UWB radar technique has greater capability to detect and locate infected tissue in the human body than microwave tomography. Microwave tomography is much more complex.

9.3 ANTENNAS FOR MICROWAVE IMAGING

9.3.1 Medical Imaging Applications

The most important component of microwave imaging is the antenna. Different challenges in designing a suitable antenna for microwave imaging, such as compact size; large bandwidth; moderate to high gain; high efficiency; very low to minimal distortion in performance, especially in the time domain; complexity; and cost, should also be considered. An ideal breast cancer detection antenna should be easily available. Again, the placement of the antenna is a challenging task. For example, for near-field applications, the antenna needs to be placed very close to the tissue, which could result in a reflection that occurs at the air–skin interface. Therefore, various types of antennas are used for different types of microwave imaging.

First we will discuss numerous types of antennas used for microwave breast cancer detection.

9.3.1.1 Monopole Antennas

Monopole antennas are widely used for microwave imaging purposes due to their large bandwidth; smaller, flexible size; and easily available nature. A UWB monopole antenna has the advantages of large bandwidth, high data rate, low cost, high resolution, low power requirement, reasonable gain, and high resistance to interference. Those promising characteristics make UWB monopole antennas an attractive choice for microwave imaging applications.

Over the last decade, numerous studies have been conducted on UWB monopole antennas. In [4], a compact UWB monopole antenna is designed with 136.5% fractional bandwidth for in-body microwave imaging applications. The antenna shows 6.1 dBi peak gain with 89% efficiency over the bandwidth. The antenna is an efficient candidate for circular cylindrical microwave imaging applications. Another body-centric imaging application is indicated in [5], where a UWB slotted monopole antenna is used. In this paper, the antenna is placed in close proximity to phantom liquid mimicking

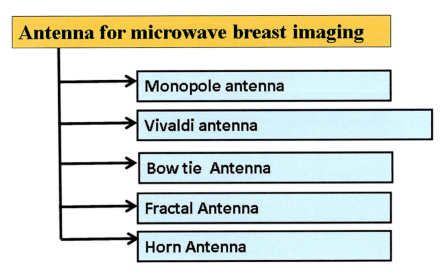

Figure 9.2 Different antennas used for microwave breast imaging.

healthy breast tissue and malignant tissue for practically verifying antenna performance. The experiment showed that the antenna is robust. [6] shows a modified octagonal UWB monopole antenna for investigation of specific absorption rate in female granular tissue in the breast. They determined SAR values in skin, fatty, glandular, and tumor tissues and compared them. The SAR values of those tissues changed or increased when a tumor was introduced in the breast tissue. However when tumor size changed, then the average SAR value from the layers was reduced. The antenna operated in the frequency range of 3–15 GHz. In [7], a UWB monopole antenna with efficiency of 86% operated in the range of 2.9–15 GHz for microwave imaging. The antenna exhibited a peak gain of 6 dBi. The antenna behaved well in the time domain measurement, as two similar antennas were placed side by side and face to face 250 mm from each other. The fidelity factor of the antennas was 0.91 and 0.84, respectively, in face-to-face and side-by-side positions. The proposed antenna is a good candidate for a microwave imaging system.

9.3.1.2 Vivaldi Antennas

For medical applications of microwave imaging, the main characteristic of the antenna should be a highly directional pattern, wide bandwidth, and enhanced gain characteristic. Vivaldi antennas have the inherent broadband characteristic with high gain and directivity. Also, the low weight and low profile quality make Vivaldi antennas preferable for microwave imaging. Much research is ongoing on Vivaldi antennas, and several modifications have been done to them. As a result, the antipodal Vivaldi antenna has been developed. [8] introduces a compact UWB tapered slot antenna for microwave imaging application purposes. To test that the antenna can transmit and receive pulses without any distortion, two identical antennas are placed at 45 cm distance. The received pulse is scaled, and it is observed that 3 dB width of the received pulse is equal to the original pulse. The distortion is less than 0.1 peak value compared to the original.

A high-gain Vivaldi antenna [9] is designed with a novel combination of corrugation and grating elements for radar and microwave imaging. Due to this combination of both corrugation and grating elements, the gain and front-to-back ratio improved significantly, almost 10 dB over 3–8 GHz. It shows a perfectly stable unidirectional radiation pattern and peak realized gain of more than 5 dBi. In the time domain characteristic, the antenna exhibits good pulse handling capability with nearly flat group delay, which indicates it is a superior candidate for microwave imaging. An antenna with the ability to send short EM pulses into the near field with low distortion and low loss with a high directional manner is designed in [10] for microwave near-field breast cancer imaging. The directivity of the

Antennas for Microwave Imaging 207

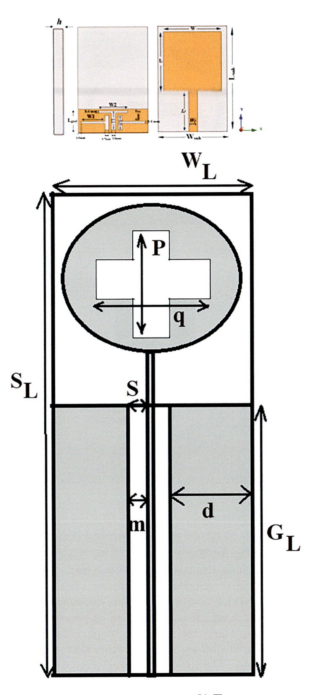

Figure 9.3 Monopole antenna for microwave imaging [4–7].

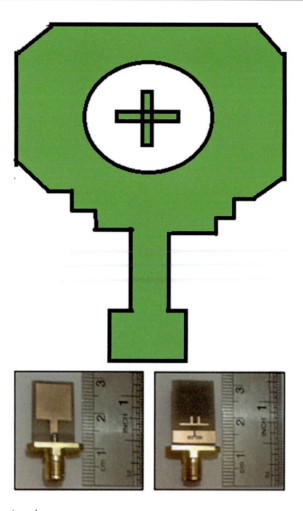

Figure 9.3 Continued

antenna is improved by adding a high dielectric constant material called a director. Figure 9.4c indicates that directivity is improved at 12 GHz with the director compared to without the director. The fidelity factor of the antenna is nearly 1, which is appropriate for microwave imaging applications. In [11], a corrugated Vivaldi antenna is developed for microwave imaging applications. Here, the authors compare the corrugated design with that of a generic Vivaldi antenna and show that the corrugated design has better performance, like higher gain, broader bandwidth, and reduction in the side lobes.

Antennas for Microwave Imaging 209

Table 9.2 Monopole Antennas for Microwave Imaging

Ref.	Dimensions (mm³)	Dielectric Material	Operating Bandwidth (GHz)	Antenna Design Type	Gain (dBi)	Image Reconstruction Algorithm
[4]	12 × 18 × 0.8	FR4 $\varepsilon r = 4.3$	2.96–15.8	Printed monopole	6.1	N/A
[5]	33.14 × 14.9 × 0.84	FR4 $\varepsilon r = 4.3$	3.1–10.6	Slotted UWB monopole	4.74	Yes (delay-multiply-sum and filtered delay and sum)
[6]	27 × 29 × 1.6	FR4 $\varepsilon r = 4.4$	3–15	Monopole octagonal UWB	–	N/A
[7]	12 × 18 × 1.6	FR4 $\varepsilon r = 4.4$	2.9–15	Square microstrip monopole antenna	6	N/A

9.3.1.3 Bow-Tie Antennas

Printed planar antennas gained more attention due to their characteristics of being easily mountable on planar surfaces and simple low-cost manufacture. Bow-tie antennas, which are the planar version of biconical antennas, can be used for microwave medical applications, with their advantages of compact size, low profile, good time domain, and broadband frequency domain radiation characteristics. Bow-tie antennas can be mainly used for ground-penetrating radar (GPR) application purposes. In some cases, they are also used in medical imaging. In [12], a UWB bow-tie antenna is used for tumor detection purposes in a radar-based microwave imaging system. The antenna performed efficiently at 1–8 GHz with good impedance matching at 40 mm depth inside the breast phantom.

A rounded bow-tie antenna [13] was fabricated on an FR4 substrate for microwave imaging applications. The antenna is compact in size so that it can be easily mounted in a scanning arrangement, efficiently optimize signal attenuation and image resolution, and receive a scattered signal from a small target. In [14], a miniaturized bow-tie antenna is sketched for medical imaging using microwave tomography. The measurement of this antenna can be done in two steps. The antenna is immersed in 60% PEG solution which mimics the dielectric properties of human tissue, and then it is applied directly to human skin. The result is matched in both cases to prove the antenna has high potential to be implemented in a microwave tomography setup. A double-crossed planar bow-tie [15] on a hemispherical lens antenna is introduced at terahertz frequency for imaging applications. To reduce

210 Printed Antennas for Wireless Communication and Healthcare

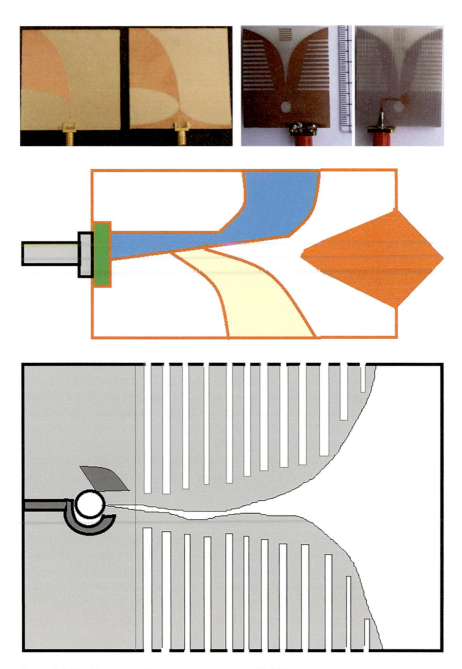

Figure 9.4 Vivaldi antenna for microwave imaging [8–11].

Antennas for Microwave Imaging   211

Table 9.3 Vivaldi Antennas for Microwave Imaging

Ref.	Dimensions (mm³)	Dielectric Material	Operating Bandwidth (GHz)	Antenna Design Type	Gain (dBi)	Image Reconstruction Algorithm
[8]	50 × 50 × 0.64	Rogers RT6010LM	2.75–11	Circular cylindrical	3.5–9.4	NA
[9]	40 × 45 × 0.8	FR4	2.9–12	Tapered corrugated and grating Tapered Vivaldi	5	NA
[10]	80 × 44 × 9.2	RT/Duroid 6002	2–16	Antipodal Vivaldi with dielectric director	–	NA
[11]	62 × 50 × 1.52	Taconic RF-35	1.96–8.61	Corrugated antipodal Vivaldi	5.6–10.4	S-parameter based linear sampling method (SLSM)

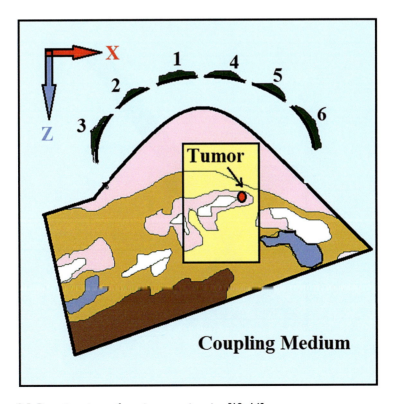

Figure 9.5 Bow-tie antenna for microwave imaging [12–16].

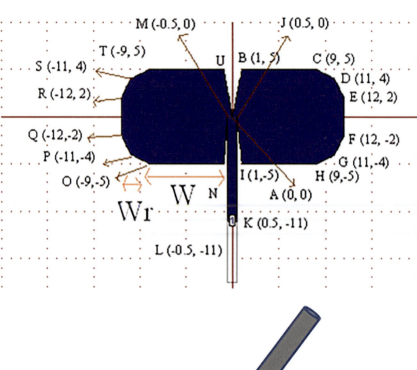

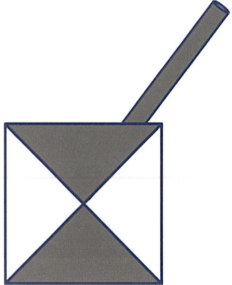

Figure 9.5 Continued

Antennas for Microwave Imaging 213

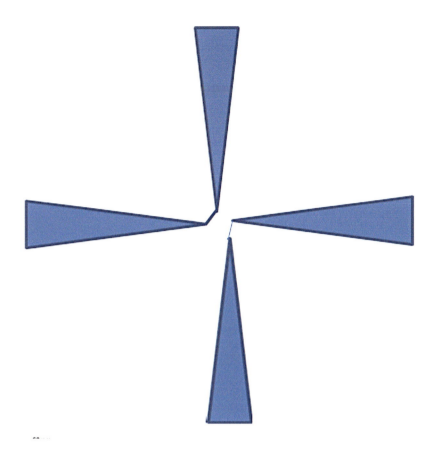

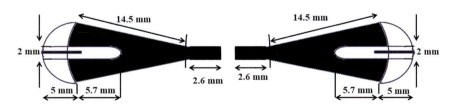

Figure 9.5 Continued

214 Printed Antennas for Wireless Communication and Healthcare

Table 9.4 Bow-Tie Antennas for Microwave Imaging

Ref.	Dimensions (mm³)	Dielectric Material	Operating Bandwidth (GHz)	Antenna Design Type	Gain (dBi)	Image Reconstruction Algorithm
[12]	26 × 40	–	1–8	UWB bow-tie	–	Delay and sum (DAS)
[13]	10 × 11 × 1.6	FR4	5.5–6.2	Rounded bow-tie	3–5	NA
[14]	30 × 30 × 1.5	Rogers 6010	0.5–1.5	Wideband bow-tie	–	NA
[15]	–	Silicon	850–1280	Double crossed bow-tie	32.26	NA
[16]	16.9 × 16.9 × 1.6	RT/ duroid 5870	3.1–40	Printed slot loaded bow-tie	7.1	NA

the side lobe level (SLL), the modification of the bow-tie antenna is done by the double-crossed technique. Two bow-tie antennas cross each other with different arm lengths. The side lobe levels at the E- and H-plane are reduced at −11.3 and −11.7 dB, respectively. Another printed modified bow-tie antenna with rounded T-shaped slots [16] loaded through a microstrip balun is implemented. The smooth impedance matching performance gives the antenna super-wideband radiation characteristics over 3.1–40 GHz with a gain of 7.1 dBi. The measured far-field radiation pattern and super-wideband characteristics of the antenna play an important role in microwave imaging techniques.

9.3.1.4 Fractal Antennas

Fractal antennas are used to improve gain, directivity, and squint, in addition to decreasing the side lobe level and increasing the main lobe. Fractal antennas have two properties: being self similar and space filling. The self-similar property gives the multiband technique, and space filling provides miniaturization. By utilizing these two properties, a fern fractal leaf-inspired wideband antipodal Vivaldi antenna [17] is applied in microwave imaging applications. The antenna displays almost 175% fractional bandwidth (1.3–20 GHz), with a high directive gain of 10 dBi and a high fidelity factor of 90%.

Another AVA of the palm tree class with a Koch square fractal [18] is proposed for near-field microwave imaging applications. The antenna provides 8.41 dB of gain at 4 GHz, 11.7 dB side lobe levels, and 0.10° squints. These types of antennas are able to detect lung cancer and lung infection caused by SARS-CoV-2, the virus caused by COVID-19. A compact square modified

Antennas for Microwave Imaging 215

fractal antenna [19] is introduced with high resolution for microwave imaging. A UWB fractal antenna with a high fidelity factor is used for microwave imaging purposes [20]. The antenna produces fidelity factors of 73%, 76%, 81%, and 83% of probes at 0°, 30°, 45°, and 90°, respectively, at 1 m distance. The antenna also shows a fidelity factor of 0.80 for face to face and 0.75 for side by side. The antenna covers the bandwidth of 3.1 to 12 GHz with a stable gain and good time domain performances.

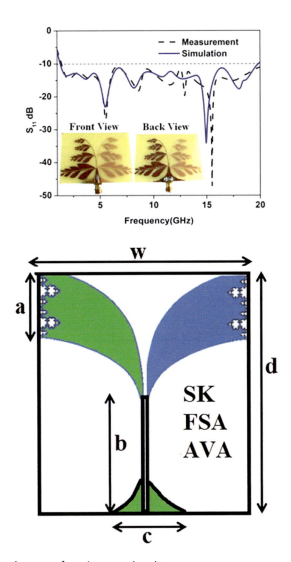

Figure 9.6 Fractal antenna for microwave imaging.

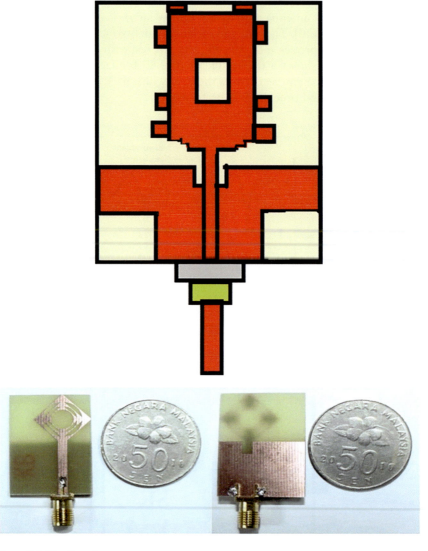

Figure 9.6 Continued

9.3.1.5 Horn Antenna

Horn antennas are used today for breast cancer detection and tracking. Horn antennas give better noise performance in comparison to the widely used Vivaldi antenna. In [21], the fact that the double ridge horn antenna gives a better response with respect to TFR, SNR, and so on in comparison

Antennas for Microwave Imaging 217

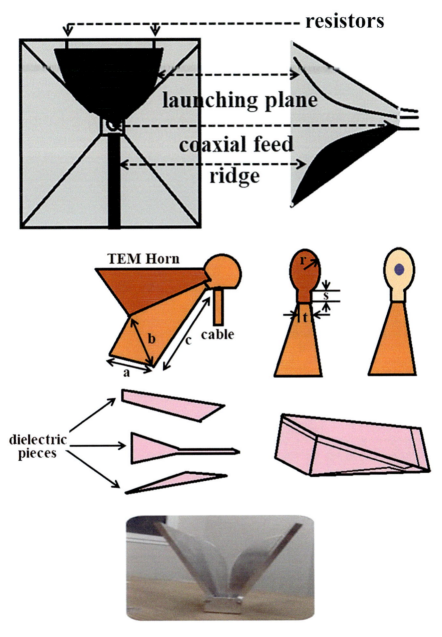

Figure 9.7 Horn antenna for microwave imaging [22–24].

218 Printed Antennas for Wireless Communication and Healthcare

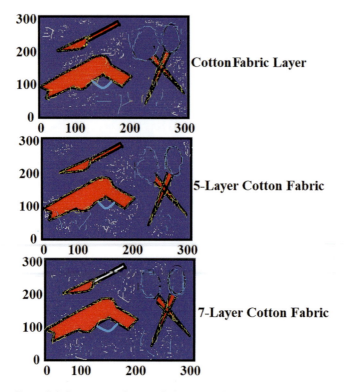

Figure 9.8 Detection of concealed weapon through microwave imaging [25].

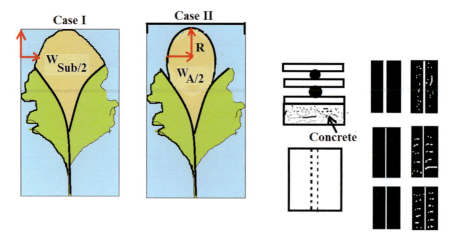

Figure 9.9 Structural health monitoring through microwave imaging [27].

Antennas for Microwave Imaging   219

to a Vivaldi antenna with same frequency response and phantom is explored. These performances of SNR and TFR allow the formation of images.

In [22], a pyramidal horn is designed with a rigid curved launching plane terminated with resistors. The antenna evinces a very high fidelity factor, almost 0.92–0.96 in the range of 1–11 GHz, for both transmission and reception mode. The distortion is also very low, which makes the antenna an excellent candidate for a variety of microwave imaging radar applications. Near-field imaging based on aperture raster scanning is done by a TEM horn antenna in [23]. The arrangements for this setup consist of two antennas aligned along each other's bore sight and moving together to scan two parallel apertures. For this type of imaging, the advantages are that no coupling liquid is required, coupling between antennas is optimal, and image generation is simple. Recently a novel and compact double-ridged horn antenna

Table 9.5 Fractal Antennas for Microwave Imaging

Ref.	Dimensions (mm³)	Dielectric Material	Operating Bandwidth (GHz)	Antenna Design Type	Gain (dBi)	Image Reconstruction Algorithm
[17]	50.8 × 62 × 0.8	FR4	1.3–20	Fern fractal leaf inspired Vivaldi	10	NA
[18]	150 × 113.71 × 1.6	FR4	1.5–5	Koch square fractal AVA	9.23	NA
[19]	39 × 39 × 1.65	FR4	2.2–11	Modified square UWB	–	NA
[20]	0.24 λ × 0.33 λ × 1.6	FR4	3.1–12	UWB fractal	2.75–3.54	NA

Table 9.6 Horn Antennas for Microwave Imaging

Ref.	Dimensions (mm³)	Fidelity Factor	Operating Bandwidth (GHz)	Antenna design type	Gain (dBi)/ Efficiency	Image reconstruction algorithm
[22]	25 × 20 × 2	0.92–0.96	1–11	Rigid curved pyramidal horn	–	NA
[23]	74 × 19 × 30	0.88	3–11	TEM horn antenna	40%	Blind de-convolution algo
[24]	151 × 108 × 146.6	–	1–9	Double-rigged horn		NA

220   Printed Antennas for Wireless Communication and Healthcare

(DRH) is illustrated for microwave imaging [24]. The antenna has a 30% size reduction in comparison to commercially available DRH antennas.

9.3.2 Concealed Weapon Detection Applications

Due to the increase of threats from national and international terrorists, the security should be tightened, especially in vulnerable spots like airports and railway stations. For this, a reliable and reasonable imaging or detection procedure is required. Many systems are available on the market, such as X-rays and metal detectors. X-ray has high resolution capability but creates high radiation, which is invasive for human beings, and metal detectors only reveal metal weapons like metal guns or knives. The effectiveness of metal detectors can vary with their quality, quantity, orientation, and type of metals. It can't differentiate between other items such as keys, belts, and glasses. It is also unable to detect modern threats like plastic and ceramic handguns and the most dangerous items, like liquid explosives like nitroglycerine; acetone peroxide; and plastic explosives like C-4 and RDX. Microwave and millimeter-wave imaging systems have a wide span to detect a variety of threatening hidden items with appreciable image resolution. It has the special capability of penetrating cloth, materials, liquids, and plastic items in the lower frequencies without any hazardous radiation. This particular characteristic makes it efficient in detection and location of concealed weapons. It mainly focuses on low atmosphere attenuation bands, which are called window bands, such as 35, 60, 95, 140, and 220 GHz. The 60-GHz band is the most popular band for security personnel surveillance systems. The 60-GHz band can detect concealed weapons behind different materials such as layers of most commonly worn fabrics, plywood sheets of different thicknesses, objects behind windows, buildings, doors, and luggage. The 60-GHz frequency can even detect four different types of explosive liquids behind clothes, plywood, and plastic materials. In [25], the authors worked with metallic and non-metallic weapons and suspected liquids hidden in clothes, plastic, and plywood with a high-gain low-profile multi-sine corrugated AFTSA antenna sensor. The authors targeted three different types of weapon, a handgun wrapped with aluminum tape; a knife with a ceramic body with a plastic handle, which is impossible to detect with a normal detector; and scissors made of stainless steel with plastic handles. The three weapons were placed in different layers of cotton fabric, plywood, and plastic sheets.

In [26], a spare aperture MIMO SAR–based UWB imaging system is designed for concealed weapon detection. Here a high-resolution imaging system based on the combination of ultra-wideband transmission, multiple-input–multi-path output array, and synthetic aperture radar is illustrated. By combining a UWB MIMO array with SAR, a high-resolution 3D volumetric imaging system is prepared for detection of concealed weapons. For MIMO-SAR application, a mannequin 1.8 m high and 0.5 m wide covered with aluminum foil is used in a standing position facing the aperture plane. Weapons like a revolver and knife are attached at the wrist and leg of the mannequin. The authors compare experiments for both a SAR-based system and MIMO-SAR system and reveal that the MIMO-SAR system obtained far better results than the SAR 2D system.

9.3.3 Structural Health Monitoring

In [27], an antipodal Vivaldi antenna with a sun-shaped configuration is designed for microwave and millimeter-wave imaging applications for structural health monitoring. Radiation characteristics, like-extended low end of frequency, increased antenna gain at lower frequencies, high front to back ratio, high gain at higher frequencies, low side lobe and cross-polarization level, and narrow half power beam width are improved by introducing a half electrical shaped dielectric lens. The antenna covers a bandwidth of 5–50 GHz. For inspection of structural health, rubber discs are embedded in construction foam at different depths.

Another multi-slot rectangular UWB antenna with a compact size of 18×10 mm^2 is used for structural health monitoring of wind turbine blades [28]. The antenna specifies a broad bandwidth of 9.2–21.4 GHz with a peak gain of 3.79–3.63 dBi, which is very helpful for health monitoring of wind turbine blades.

9.4 SUMMARY

This chapter details about the pros and cons of microwave imaging technique along with its antenna theory. The essence of the microwave imaging in connection with bio-medical engineering has been discussed here. The types of imaging processes and its antennas have been highlighted. Monopole, Vivaldi, antipodal Vivaldi, Bow-tie, fractal and horn are some prime candidates in its antenna field. Various reported structures have been compared quantitatively, which acts as a ready reference for the beginners apart from the bio-medical field, usage of this imaging technique in strategic and commercial domain is also demonstrated with real life problem.

REFERENCES

[1] E. C. Fear, X. Li, S. C. Hagness, and M. A. Stuchly, "Confocal Microwave Imaging for Breast Cancer Detection: Localization of Tumors in Three Dimensions," *IEEE Transactions on Biomedical Engineering*, Vol. 49, pp. 812–822, 2002.

[2] X. Li, and S. C. Hagness, "A Confocal Microwave Imaging Algorithm for Breast Cancer Detection," *IEEE Microwave and Wireless Components Letters*, Vol. 11, no. 3, pp. 130–132, 2001.

[3] S. C. Hagness, A. Taflove, and J. E. Bridges, "Two-Dimensional FDTD Analysis of a Pulsed Microwave Confocal System for Breast Cancer Detection: fixed-Focus and Antenna-Array Sensors," *IEEE Transactions on Biomedical Engineering*, Vol. 45, pp. 1470–1479, 1998.

[4] A. Aref, P. Abbas, E. Homauon, and A. Mousa, "A Compact UWB Printed Antenna with Bandwidth Enhancement for in Body Microwave Imaging Applications," *Progress in Electromagnetics Research C*, Vol. 55, pp. 149–157, 2014, doi: 10.2528/PIERC14111405.

[5] M. Danjuma, M. O. Akinsolu, C. H. See, R. A. Abd-Alhameed, and B. Liu, "Design and Optimization of a Slotted Monopole Antenna for Ultra-Wide Band Body Centric Imaging Applications," *IEEE Journal of Electromagnetics, RF and Microwaves in Medicine and Biology*, Vol. 4, no. 2, pp. 140–147, 2020, doi: 10.1109/JERM.2020.2984910.

[6] S. Subramanian, B. Sundarambal, and D. Nirmal, "Investigation on Simulation-Based Specific Absorption Rate in Ultra-Wideband Antenna for Breast Cancer Detection," *IEEE Sensors Journal*, Vol. 18, no. 24, pp. 10002–10009, 2018, doi: 10.1109/JSEN.2018.2875621.

[7] N. Ojaroudi, and N. Ghadimi, "Omnidirectional Microstrip Monopole Antenna Design for Use in Microwave Imaging System," *Microwave and Optical Technology Letter*, Vol. 57, no. 2, pp. 395–401, 2015.

[8] A. M. Abbosh, H. K. Kan, and M. E. Bialkowski, "Compact Ultra-Wideband Planar Tapered Slot Antenna for Use in a Microwave Imaging System," *Microwave and Optical Technology Letter*, Vol. 48, no. 11, pp. 2212–2216, 2006.

[9] G. K. Pandey, and M. K. Meshram, "A Printed High Gain UWB Vivaldi Antenna Design Using Tapered Corrugation and Grating Elements," *International Journal of RF and Microwave Computer Aided Engineering*, Vol. 25, no. 7, pp. 610–618, 2015.

[10] J. Bourqui, M. Okoniewski, and E. C. Fear, "Balanced Antipodal Vivaldi Antenna with Dielectric Director for Near-Field Microwave Imaging," *IEEE Transactions on Antennas and Propagation*, Vol. 58, no. 7, pp. 2318–2326, 2010, doi: 10.1109/TAP.2010.2048844.

[11] M. Abbak, M. N. Akıncı, M. Çayören, and İ. Akduman, "Experimental Microwave Imaging with a Novel Corrugated Vivaldi Antenna," *IEEE Transactions on Antennas and Propagation*, Vol. 65, no. 6, pp. 3302–3307, 2017, doi: 10.1109/TAP.2017.2670228.

[12] İ. Ünal, B. Türetken, and Y. Çotur, "Microwave Imaging of Breast Cancer Tumor Inside Voxel-Based Breast Phantom Using Conformal Antennas," *2014 XXXIth URSI General Assembly and Scientific Symposium* (URSI GASS), pp. 1–4, 2014, doi: 10.1109/URSIGASS.2014.6930132.

[13] E. Rufus, Z. C. Alex, and P. V. Chaitanya, "A Modified Bow-Tie Antenna for Microwave Imaging Applications," *Journals of Microwave, Optoelectronics and Electromagnetic Application*, Vol. 7, no. 2, 2008.

[14] M. Jalilvand, C. Vasanelli, C. Wu, J. Kowalewski, and T. Zwick, "On the Evaluation of a Proposed Bowtie Antenna for Microwave Tomography," *The 8th European Conference on Antennas and Propagation* (EuCAP 2014), pp. 2790–2794, 2014, doi: 10.1109/EuCAP.2014.6902405.

[15] A. P. Aji, C. Apriono, F. Y. Zulkifli, and E. T. Rahardjo, "Double Crossed Planar Bow-Tie on a Lens Antenna at Terahertz Frequency for Imaging Application," *Progress in Electromagnetics Research Symposium* (PIERS | Toyama), 2018.

[16] O. Yurduseven, D. Smith, and M. Elsdon, "Printed Slot Loaded Bow-Tie Antenna with Super Wideband Radiation Characteristics for Imaging Applications," *IEEE Transactions on Antennas and Propagation*, Vol. 61, no. 12, pp. 6206–6210, 2013, doi: 10.1109/TAP.2013.2281353.

[17] B. Biswas, R. Ghatak, and D. R. Poddar, "A Fern Fractal Leaf Inspired Wideband Antipodal Vivaldi Antenna for Microwave Imaging System," *IEEE Transactions on Antennas and Propagation*, Vol. 65, no. 11, pp. 6126–6129, 2017, doi: 10.1109/TAP.2017.2748361.

[18] R. E. Figueredo, A. M. de Oliveira, N. Nurhayati, A. M. de O. Neto, I. C. Nogueira, J. F. Justo, M. B. Perotoni, and A. de Carvalho Jr, "A Vivaldi Antenna Palm Tree Class with Koch Square Fractal Slot Edge for Near-Field Microwave Biomedical Imaging Applications," *2020 Third International Conference on Vocational Education and Electrical Engineering* (ICVEE), pp. 1–6, 2020, doi: 10.1109/ICVEE50212.2020.9243220.

[19] H. M. Q. Rasheda et al., "Design of UWB Antenna for Microwave Imaging using Modified Fractal Structure," *2021 International Congress of Advanced Technology and Engineering* (ICOTEN), pp. 1–4, 2021, doi: 10.1109/ICOTEN52080.2021.9493440.

[20] M. M. Islam et al., "Microstrip Line Fed Fractal Antenna with a High Fidelity Factor for UWB Imaging Application," *Microwave and Optical Technology Letters*, Vol. 57, no. 11, pp. 2580–2585, 2015.

[21] M. Solis Nepote, D. R. Herrera, D. F. Tapia, S. Latif, and S. Pistorius, "A Comparison Study between Horn and Vivaldi Antennas for 1.5–6 GHz Breast Microwave Radar Imaging," *The 8th European Conference on Antennas and Propagation* (EuCAP 2014), pp. 59–62, 2014, doi: 10.1109/EuCAP.2014.6901692.

[22] S. C. H. Xu Li, M. K. Choi, and D. W. van der Weide, "Numerical and Experimental Investigation of an Ultrawideband Ridged Pyramidal Horn Antenna with Curved Launching Plane for Pulse Radiation," *IEEE Antennas and Wireless Propagation Letters*, Vol. 2, pp. 259–262, 2003, doi: 10.1109/LAWP.2003.820708.

[23] R. K. Amineh, M. Ravan, A. Trehan, and N. K. Nikolova, "Near-Field Microwave Imaging based on Aperture Raster Scanning with TEM Horn Antennas," *IEEE Transactions on Antennas and Propagation*, Vol. 59, no. 3, pp. 928–940, 2011, doi: 10.1109/TAP.2010.2103009.

[24] S. Diana, D. Brizi, C. Ciampalini, G. Nenna, and A. Monorchio, "A Compact Double-Ridged Horn Antenna for Ultra-Wide Band Microwave Imaging," *IEEE Open Journal of Antennas and Propagation*, Vol. 2, pp. 738–745, 2021, doi: 10.1109/OJAP.2021.3089028.

[25] Z. Briqech, S. Gupta, A. A. Beltay, A. Elboushi, A. R. Sebak, and T. A. Denidni, "57–64 GHz Imaging/Detection Sensor—Part II: Experiments on Concealed Weapons and Threatening Materials Detection," *IEEE Sensors Journal*, Vol. 20, no. 18, pp. 10833–10840, 2020, doi: 10.1109/JSEN.2020.2997293.

[26] X. Zhuge, and A. G. Yarovoy, "A Sparse Aperture MIMO-SAR-Based UWB Imaging System for Concealed Weapon Detection," *IEEE Transactions on*

Geoscience and Remote Sensing, Vol. 49, no. 1, pp. 509–518, 2011, doi: 10.1109/TGRS.2010.2053038.

[27] M. Moosazadeh, S. Kharkovsky, J. T. Case, and B. Samali, "Improved Radiation Characteristics of Small Antipodal Vivaldi Antenna for Microwave and Milli-meter-Wave Imaging Applications," *IEEE Antennas and Wireless Propagation Letters*, Vol. 16, pp. 1961–1964, 2017, doi: 10.1109/LAWP.2017.2690441.

[28] P. Muthusamy, and P. V. Durairaj, "An Overview of Microwave UWB Antenna for Structural Health Monitoring of Wind Turbine Blades," *Optimal Design and Analysis, Instrumentation Mesure Metrologie*, Vol. 18, no. 1, pp. 75–81, 2019, doi: 10.18280/i2m.180112

Problems

1. Why is microwave preferred for imaging purposes?
2. Name a few important applications of microwave imaging.
3. What are the different types of microwave imaging?
4. What are the different types of antennas used for microwave imaging applications?
5. What are the specialties of fractal antennas?
6. Why have Vivaldi antennas become so popular in microwave imaging?
7. How can microwave imaging be utilized to detect concealed weapons?
8. How the microwave imaging technique can be implemented for health monitoring of any machine in industry?
9. How is microwave imaging treated as a better/the best option compared to its counterparts?

Chapter 10

Rectennas

A New Frontier for Future Wireless Communication

10.1 INTRODUCTION

With the rapid growth in IC and embedded system technology, self-supporting low-power devices suitable for the IoT, RFID systems, smart cities, wearable electronics, wireless sensor networks, and so on have attracted notable attention in the last few years. Presently, the power requirements of these remote devices are generally met using standard cells or batteries. Once exhausted, these DC power sources are immediately replaced with new ones. This itself is a tedious job in addition to being hazardous to the environment. Hence, there is a need for a sustainable or green solution. Energy harvesting (EH), wherein abundantly available energy is absorbed by several means and thereafter converted into a DC power source for powering devices, is the most suitable for this issue.

There are various sources of ambient energy, like the sun, wind, tides, piezoelectric vibrations, and EM waves. Among them, the energy associated with EM waves or RF energy source is the most easily available, and it is abundant due to the explosive growth of mobile communication, WiFi, TV users, and so on in the modern era. Hence, it can be used as a primitive resource for EH systems in the current scenario. Table 10.1 summarizes comparative studies of possible sources of energy harvesting in terms of power density, harvesting technique, advantages, and inherent disadvantages.

Radio frequency energy harvesting and wireless power transferring methods gained wide recognition over a decade ago in the process of enabling battery-free wireless networks. Rectifying antennas, or rectennas, are the backbone of such systems. The performance of the whole system greatly depends on the quality of the rectenna developed for it. This kind of antenna picks up electromagnetic waves from the surroundings. Whenever it receives a signal, it generates oscillating charges that move through attached fluctuations to a direct electric current. The throughput of the system solely depends upon the AC-to-DC conversion.

The main concept of wireless power transferring (WPT) is not new. It is dates back to 1890s, when Nikola Tesla originally proposed it for high-power applications. However, practical applications started much later due

DOI: 10.1201/9781003389859-10

225

226 Printed Antennas for Wireless Communication and Healthcare

Table 10.1 Overview of Probable Alternative Sources of Energy to Replace Batteries/Cells

Source	Power Density	Harvesting Technique	Advantages	Disadvantages
Sun [1]	Indoor: 10 µW/cm² Outdoor: 10 mW/cm²	Photovoltaic	High power density	Not always available Expensive
Thermal [1]	Human: 30 µW/cm² Industrial: 1–10 mW/cm²	Thermoelectric Pyro-electric	High power density	Not always available Excess heat
Vibration [1]	Human: 4 µW/cm² Industrial: 100 µW/cm²	Piezoelectric Electrostatic Electromagnetic	High power density	Not always available Material science limitations
RF [1]	GSM: 0.1 µW/cm² WiFi: 1 mW/cm²	Antenna	Always available Implantable	Low density Efficiency inversely proportional to distance

to various technical bottlenecks. The term "Rectenna" refers an antenna along with some rectifying circuit, which can be used for harvesting the RF energy from the surroundings [2].

As time progressed towards a revolution in wireless communication, the power consumption of semi-conductor devices and wireless sensors nodes continuously went down. It became more feasible to power sensor nodes using ambient radio frequency energy harvesters. Such a system mainly consists of an RF harvesting front end, DC power source, storage unit, and low-power microprocessor and transceiver. Figure 10.1 depicts the architecture of an RF energy harvester. The efficiency of such a system depends upon the performance of individual blocks, such as the antenna, rectifier, and power management unit. Figure 10.2 shows the possible sources of RF energy in the ambient environment. Table 10.2 summarizes a variety of rectenna circuits developed in recent times, whereas Table 10.3 briefly outlines various WPT techniques along with their performance metrics. Figure 10.3 illustrates WPT techniques.

In the subsequent sections, a detailed analysis of rectenna circuits is carried out. A recent literature survey enriches the chapter, with the latest state-of-the art architectures and their inherent merits and demerits.

10.2 PERFORMANCE METRICS OF RF HARVESTERS

There are various parameters that directly or indirectly decide the performance of the whole energy harvesting system, and they need to be evaluated rigorously. These are: efficiency, sensitivity, operating distance, output power, maturity of fabrication/manufacturing process, and so on.

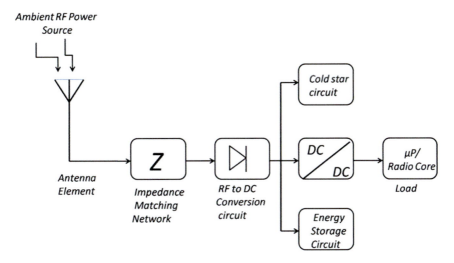

Figure 10.1 Generalized architecture of RF energy harvester.

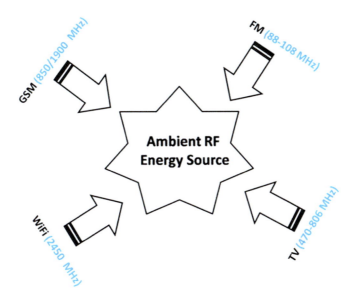

Figure 10.2 Available RF energy sources in ambient environment.

10.2.1 Range of Operation

The range of operation is strongly dependent on the frequency of the operation. Transmission of the higher-frequency side is affected badly by various loss factors associated with the atmosphere. At low frequencies, the signal

228   Printed Antennas for Wireless Communication and Healthcare

Table 10.2 A Comparison of Rectenna Circuits

Literature	Antenna	Matching Technique	Frequency Bands (GHz)
2018[3]	Narrowband patch	Single-band tapered line	2.45
2018[4]	Broadband slot, single band slot	Tee or Pi network	0.9, 2, 2.55
2017[5]	High-Z, multiband dipole	N/A	0.95, 1.85–2.4
2016[6]	Frequency-independent log periodic	Transmission line method	0.55, 0.75, 0.9, 1.8, 2.3
2016[7]	High-Z dipole	N/A	0.55
2014[8]	High-Q loop	Weighted capacitor bank	0.868
2013[9]	Broadband Yagi-Uda array	8th-order LC network	1.8, 2.1

Table 10.3 Comparison Between Various WPT Techniques

Field Region	WPT Technique	Nature of Propagation	Efficiency
Near-field	Resonant inductive coupling	Non-radiative	5.81–57.2% (for 16.2 to 508 KHz) [2]
	Magnetic resonance coupling	Non-radiative	30 to 90%
Far-field	RF energy transfer	Radiative	0.4%, >18.2% and >50% at −40, −20, and −5 dBm input power, respectively [2]

can easily penetrate matter. Hence, for the application of radio frequency energy harvesting, the frequency of the operation should be chosen very wisely.

10.2.2 Power Conversion Efficiency of RF to DC

This is the ratio of power applied to the load and that retrieved by the antenna element. In RF-to-DC conversion, the PCE covers the efficiency of the rectifying circuit, voltage multiplier, and storage unit. PCE can be mathematically expressed as

$$\eta_{PCE} = \frac{P_{load}}{P_{retrieved}} \tag{10.1}$$

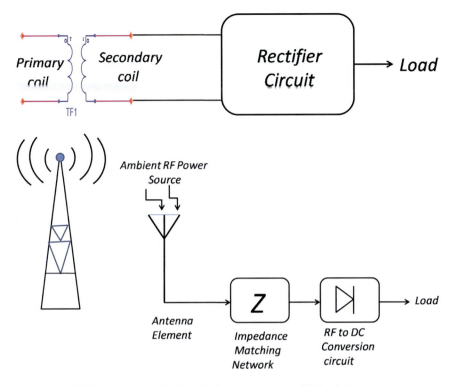

Figure 10.3 WPT techniques. (a) Near-field communication. (b) Far-field communication.

where P_{load} is the power delivered to the load, and $P_{retrieved}$ is the power retrieved at the antenna side.

In this equation, RF transmission loss is neglected.

10.2.3 Q-Factor of the Resonator

Q-factor or quality factor is defined as the ratio between the maximum energy stored by the resonator and the energy dissipated per cycle. It decides the resonance frequency as well as the operational bandwidth. It can be defined as follows:

$$Q = 2\pi \frac{\text{Maximum energy stored within resonator}}{\text{Energy dissipated per cycle}} = \frac{f_0}{BW} \quad (10.2)$$

$$\text{Capacitor Q-factor, } Q_C = \frac{1}{\omega RC} \quad (10.3)$$

Inductor Q-factor, $Q_L = \dfrac{\omega L}{R}$ (10.4)

10.2.4 Sensitivity

The sensitivity of a harvesting system is defined as the minimum amount of incident signal power that is sufficient to trigger/initiate the operation of the system. It is generally expressed in mV/dBm.

10.2.5 Output Power

The output of a harvesting system is DC power, which is further characterized by load voltage V_{DD} and current I_{DD}.

If the load is a sensor, V_{DD} is more important than I_{DD}, while in applications like electrolysis or LEDs, the main dominant parameter is I_{DD}.

10.3 DESIGN PROTOCOL OF RF ENERGY HARVESTING CIRCUITS

In previous sections, the constituent building blocks of RF energy harvesting systems are described. The overall efficiency of the system relies upon the efficiency of individual elements. Hence, optimization is essential at the design level for individual building blocks. The work flow of designing such a harvesting system is shown in Figure 10.4.

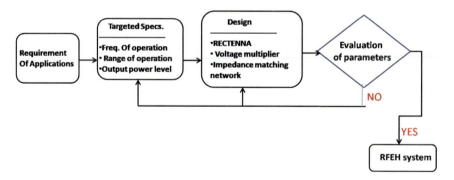

Figure 10.4 Work flow of designing an RF harvesting system.

10.4 BUILDING BLOCKS OF RECTENNAS

10.4.1 Front-End Antennas

As the electronic eye of the whole system, the antenna plays an important role in energy harvesting circuit. There are several parameters that decide the performance metrics of the antenna. These are: return loss, impedance bandwidth, gain, directivity, polarization, radiation efficiency, and so on. A literature survey reveals that people have tried single as well as multiband antennas for this purpose to achieve maximum throughput of the system. Further, the polarization features have also been altered to attain maximum efficacy of the energy harvester. As a key component of the whole system, the operating frequency of the antenna is chosen wisely so that the maximum ambient RF energy can be captured. Usually, the frequencies of operations are 915 MHz, 2.45 GHz, and 5.8 GHz, and the types of antenna elements used are microstrip patch, Yagi-Uda, dipole, aperture coupled patch, stacked patch antennas, and so on. The gain of the antenna mainly varies from 3 to 20 dBi, whereas RF-to-DC power conversion efficiency can be achieved up to 83 %. Various types of antennas used/developed for this purpose are shown in Figures 10.5 to 10.7.

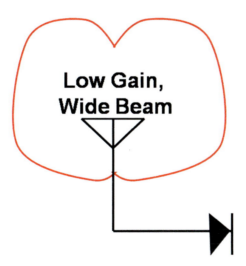

Figure 10.5 Rectenna topologies based on radiation patterns. (a) Omnidirectional antenna. (b) High-gain directional antenna. (c) Multiband high-gain narrow beams with RF combiner. (d) Multiband high-gain narrow beams with DC combiner [10].

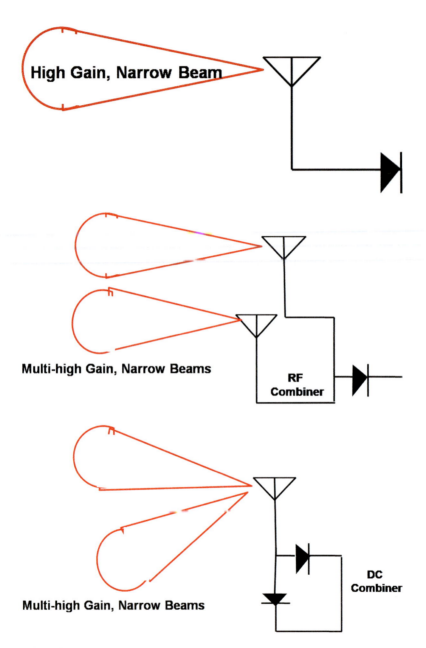

Figure 10.5 Continued

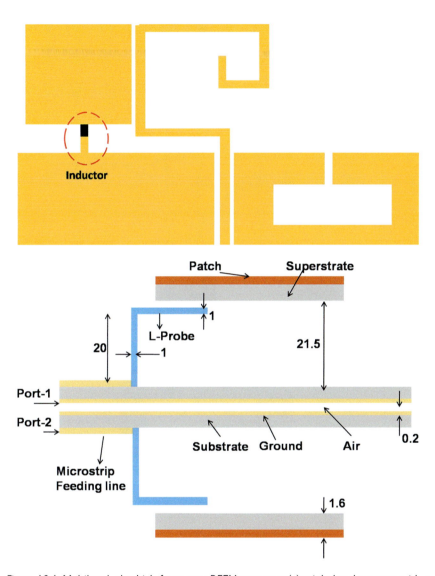

Figure 10.6 Multiband, ultrahigh-frequency RFEH antennas: (a) triple-band antenna with a lumped inductor and three radiator elements [11], (b) triple-band slotted patch [4], and (c) one-probe-fed, dual-band patch [12].

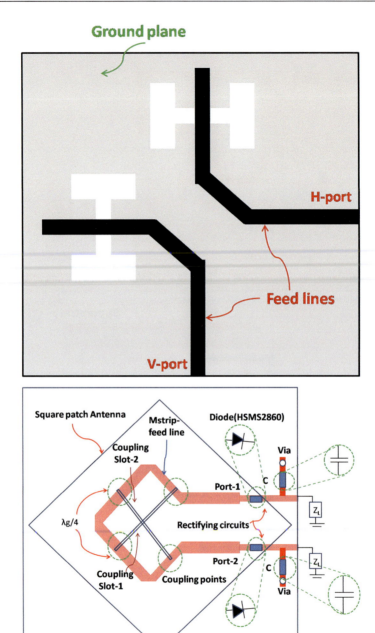

Figure 10.7 Polarization-independent rectennas: (a) dual-LP slot [13], (b) dual-CP slot [14], (c) dual-LP cross-dipole array [15], and (d) dual antennas for harvesting near-field (top) H- and (bottom) E-fields [16].

10.4.2 Impedance Matching Networks

Impedance matching networks (IMNs) play an important role in making a smooth transition between the impedance level of the antenna's input feeding network and load impedance of the rectifying circuit. Maximum power can only be transferred if there is conjugate matching of impedance at the RF level.

Several types of matching topologies are adopted as depicted in Figure 10.8. Depending upon the suitability of the application engineering, the type of matching network is selected. For multiband antenna applications, sometimes tunable IMNs are implemented instead of fixed ones. Usually, three basic configurations of matching networks are used extensively: L, T, or π. Commonly, L-type matching is used, which simplifies the design and control process. It doesn't alter the Q-factor of the circuit. However, complex design problems demand a T or pi network.

10.4.3 Rectifying Circuits

Usually, the density of the ambient RF energy/power is feeble, so processing it through the antenna and input matching network will further decrease its power level. Hence, along with RF-to-DC conversion activity, it becomes essential to have a boost or multiply the output voltage level.

Some common topologies of rectifiers are shown in Figure 10.9. Among these, the most primitive is the half-wave rectifier, which includes a single

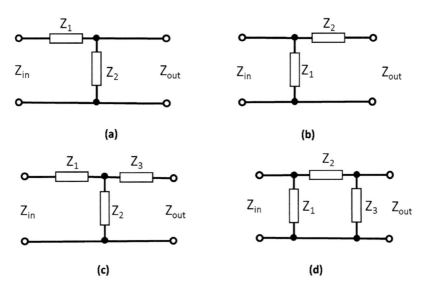

Figure 10.8 Impedance matching circuit of (a) forward L-type, (b) reverse L-type, (c) T network, and (d) π network.

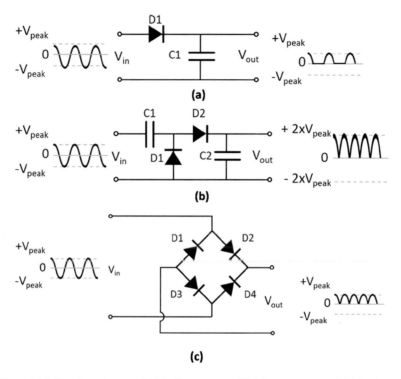

Figure 10.9 Rectifier circuits of (a) half-wave type, (b) full-wave type, and (c) bridge type.

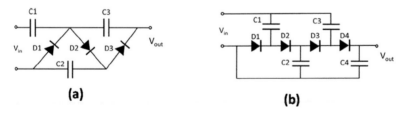

Figure 10.10 Voltage multiplier types: (a) Cockcroft-Walton and (b) Dickson.

diode. It allows only half of the input AC/RF waveform. For smoother AC-to-DC conversion, a full-wave rectifier or bridge network is implemented.

A voltage multiplier is another kind of rectifier circuit that converts and boosts AC input to DC output. There are some cases where rectified power is inadequate for the rest of the applications. Hence, the importance of a voltage multiplier circuit is realized. Basically, there are two types of voltage multipliers: Cockcroft-Walton and Dickson [1] (Figure 10.10).

10.5 PRACTICAL APPLICATIONS OF RF ENERGY HARVESTERS

10.5.1 Healthcare Devices/Biomedical Engineering

The healthcare sector or biomedical engineering demands sensors or devices with very low power needs for general operation. Radio frequency energy harvesting may be the answer for all of these devices, replacing the current DC power source, that is, the battery/cell. In a very efficient way, the RFEH technique can be an alternative to all sensors. A battery-free bio-signal processing System-on-chip (SoC) is demonstrated in [17]. The system is able to monitor various bio-signals via ECG, EEG, EMG, and so on. The overall size of the chip is 8.25 mm^2, which only consumes 19 µW to measure the heart rate. The power module of this system consists of RFEH, which can supply around 1.55 Volts (DC). Table 10.4 presents the performance of several health-monitoring SoCs.

10.5.2 Future Wireless Communication Networks

In the very near future, wireless communications will be established between various wireless sensor networks in smart cities, the IoT, and even 5G. With the rapid growth of micro- and nano-technology along with IC fabrication and embedded systems, the realization of various sensors is imminent.

In such WSNs, energy is a critical concern, along with sensing reliability, communication protocols used, and network services. Fortunately, with CMOS and MEMS technology, sensors are becoming miniaturized and need very low power for normal operation. Utilizing harvested RF energy to recharge batteries is one approach to enhance the lifespan of WSNs. [22] demonstrates an RF charger implementing a wireless harvester. The system is able to work with minimum power of 0.1 mW in the frequency range of 1.96 to 1.98 GHz to charge a 5V super-capacitor. The maximum RF-to-DC power conversion efficiency (PCE) achieved was 81% at 6 dBm (3.98 mW).

Table 10.4 Various Health-Monitoring Sensors With Power-Consuming Capabilities

Reference	Sensor	Chip Power (µW)
[18]	EEG	1
[19]	ECG	1.2
[20]	Intra-ocular pressure (IOP)	—
[21]	Neural, ECG, EMG, EEG	1

10.5.3 Explosive Detection Missions

In the near future, robots are planned to be used for this purpose. The robot has battery/cell inside it, and it requires recharging after a few hours, so recharging it without human intervention by wireless means will be a blessing for technologists. These robots also operate at larger ranges.

Near-field coupling (NFC) will be suitable up to s few feet, whereas if the range/distance is larger, WPT with the means of a microwave or laser technique will be ideal. A laser technique needs line-of-sight (LOS) communication, whereas microwave doesn't particularly need a LOS communication link. Even through physical barriers (non-metallic in nature), microwave can penetrate. The only hurdle for it is the moisture content of surroundings. If there is heavy rain or a dense forest, its power budget will be compromised.

10.5.4 Unique Solutions for Specific Military Applications

Radio-frequency energy-harvesting technology is very well suited to some of the unique challenges faced by combatants in current military vehicles, such as supplying power to individuals seated in the vehicle. Today, soldiers carry many electrical and electronic systems that provide them with advanced capabilities. However, these advances have come with a cost: it has become necessary for soldiers to carry many spare (non-rechargeable) batteries to keep these systems operational. An individual could be required to carry approximately 25 lb of battery weight to support a typical three-day mission. Hence, any opportunity to recharge the batteries or provide temporary operating power for these systems could offer major benefits by reducing the weight. With these challenges as a motivation, the possibility of transferring power from source to load using non-contact wireless means can be investigated, or RFEH using a robot remotely operated by a soldier could be very useful.

10.6 SUMMARY

In this chapter, the essence of rectenna circuits is discussed in detail in context with the rapid growth of mobile or wireless communication. The evolution of energy harvesting circuits is also highlighted here. Comparisons between various harvesting techniques have been given. Various performance metrics of harvesters are also outlined to evaluate the quality of the built prototypes. Subsequently, the building blocks of rectenna modules are explained. Antenna design is detailed with a literature survey. Keeping in view of fabrication tolerances, the design parameters of antenna architecture is finalized. Polarization of antennas plays an important role in such applications. With the next key component being the impedance matching network, its design becomes a vital point. Depending upon the system requirements, the type of matching network is selected, and finally, the rectifier circuit finishes its job to successfully convert AC to DC. The power conversion efficiency depends upon lot of factors, which have been discussed in the chapter.

Apart from design strategies and fabrication challenges, the chapter concludes with the possible application of rectenna circuits targeting multifaceted wireless communications.

REFERENCES

[1] L. G. Tran, H. K. Cha, and W. T. Park, "RF Power Harvesting: A Review on Designing Methodologies and Applications," *Micro and Nano Systems Letter*, Vol. 5, no. 14, 2017, doi: 10.1186/s40486-017-0051-0

[2] N. Shinohara, "Trends in Wireless Power Transfer: WPT Technology for Energy Harvesting, Millimeter-Wave/THz Rectennas, MIMO-WPT, and Advances in Near-Field WPT Applications," *IEEE Microwave Magazine*, Vol. 22, no. 1, pp. 46–59, 2021, doi: 10.1109/MMM.2020.3027935.

[3] S. E. Adami et al., "A Flexible 2.45-GHz Power Harvesting Wristband with Net System Output From –24.3 dBm of RF Power," *IEEE Transactions on Microwave Theory and Techniques*, Vol. 66, no. 1, pp. 380–395, 2018, doi: 10.1109/TMTT.2017.2700299.

[4] V. Palazzi et al., "A Novel Ultra-Lightweight Multiband Rectenna on Paper for RF Energy Harvesting in the Next Generation LTE Bands," *IEEE Transactions on Microwave Theory and Techniques*, Vol. 66, no. 1, pp. 366–379, 2018, doi: 10.1109/TMTT.2017.2721399.

[5] C. Song et al., "Matching Network Elimination in Broadband Rectennas for High-Efficiency Wireless Power Transfer and Energy Harvesting," *IEEE Transactions on Industrial Electronics*, Vol. 64, no. 5, pp. 3950–3961, 2017, doi: 10.1109/TIE.2016.2645505.

[6] C. Song et al., "A Novel Six-Band Dual CP Rectenna Using Improved Impedance Matching Technique for Ambient RF Energy Harvesting," *IEEE Transactions on Antennas and Propagation*, Vol. 64, no. 7, pp. 3160–3171, 2016, doi: 10.1109/TAP.2016.2565697.

[7] H. Miyagoshi, K. Noguchi, K. Itoh, and J. Ida, "High-Impedance Wideband Folded Dipole Antenna for Energy Harvesting Applications," *2014 International Symposium on Antennas and Propagation Conference Proceedings*, pp. 601–602, 2014, doi: 10.1109/ISANP.2014.7026794.

[8] M. Stoopman, S. Keyrouz, H. J. Visser, K. Philips, and W. A. Serdijn, "Co-Design of a CMOS Rectifier and Small Loop Antenna for Highly Sensitive RF Energy Harvesters," *IEEE Journal of Solid-State Circuits*, Vol. 49, no. 3, pp. 622–634, 2014, doi: 10.1109/JSSC.2014.2302793.

[9] H. Sun, Y. Guo, M. He, and Z. Zhong, "A Dual-Band Rectenna Using Broadband Yagi Antenna Array for Ambient RF Power Harvesting," *IEEE Antennas and Wireless Propagation Letters*, Vol. 12, pp. 918–921, 2013, doi: 10.1109/LAWP.2013.2272873.

[10] M. Wagih, A. S. Weddell, and S. Beeby, "Rectennas for Radio-Frequency Energy Harvesting and Wireless Power Transfer: A Review of Antenna Design [Antenna Applications Corner]," *IEEE Antennas and Propagation Magazine*, Vol. 62, no. 5, pp. 95–107, 2020, doi: 10.1109/MAP.2020.3012872.

[11] B. L. Pham, and A. Pham, "Triple Bands Antenna and High Efficiency Rectifier Design for RF Energy Harvesting at 900, 1900 and 2400 MHz," *2013 IEEE MTT-S International Microwave Symposium Digest (MTT)*, pp. 1–3, 2013, doi: 10.1109/MWSYM.2013.6697364.

[12] S. Shen, C. Chiu, and R. D. Murch, "A Dual-Port Triple-Band L-Probe Microstrip Patch Rectenna for Ambient RF Energy Harvesting," *IEEE Antennas and Wireless Propagation Letters*, Vol. 16, pp. 3071–3074, 2017, doi: 10.1109/LAWP.2017.2761397.

[13] H. Sun, and W. Geyi, "A New Rectenna with All-Polarization-Receiving Capability for Wireless Power Transmission," *IEEE Antennas and Wireless Propagation Letters*, Vol. 15, pp. 814–817, 2016, doi: 10.1109/LAWP.2015.2476345.

[14] H. Zied, C. Laurent, O. Lotfi, G. Ali, Senior Member, IEEE, and P. Odile, "A Dual Circularly Polarized 2.45-GHz Rectenna for Wireless Power Transmission," *IEEE Antennas And Wireless Propagation Letters*, Vol. 10, 2011.

[15] A. Z. Ashoor, and O. M. Ramahi, "Polarization-Independent Cross-Dipole Energy Harvesting Surface," *IEEE Transactions on Microwave Theory and Techniques*, Vol. 67, no. 3, pp. 1130–1137, 2019, doi: 10.1109/TMTT.2018.2885754.

[16] T. Lin, J. Bito, J. G. Hester, J. Kimionis, R. A. Bahr, and M. M. Tentzeris, "Ambient Energy Harvesting from Two-Way Talk Radio for On-Body Autonomous Wireless Sensing Network Using Inkjet and 3D Printing," *2017 IEEE MTT-S International Microwave Symposium (IMS)*, pp. 1034–1037, 2017, doi: 10.1109/MWSYM.2017.8058767.

[17] Y. Zhang, F. Zhang, Y. Shakhsheer, J. D. Silver, A. Klinefelter, M. Nagaraju, J. Boley, J. Pandey, A. Shrivastava, E. J. Carlson, A. Wood, B. H. Calhoun, and B. P. Otis, "A Batteryless 19 µw MICS/ISM-band Energy Harvesting Body Sensor Node SoC for ExG Applications," *IEEE Journal of Solid-State Circuits*, Vol. 48, no. 1, pp. 199–213, 2013, [6399579] doi: 10.1109/JSSC.2012.2221217

[18] Verma, Naveen et al., "A Micro-Power EEG Acquisition SoC With Integrated Feature Extraction Processor for a Chronic Seizure Detection System." *IEEE Journal of Solid-State Circuits*, Vol. 45, no. 4, pp. 804–816, 2010.

[19] H. Kım et al., "A Configurable and Low-Power Mixed Signal SoC for Portable ECG Monitoring Applications," *IEEE Transactions on Biomedical Circuits and Systems*, Vol. 8, no. 2, pp. 257–267, 2014, doi: 10.1109/TBCAS.2013.2260159.

[20] G. Chen et al., "A Cubic-Millimeter Energy-Autonomous Wireless Intraocular Pressure Monitor," *2011 IEEE International Solid-State Circuits Conference*, pp. 310–312, 2011, doi: 10.1109/ISSCC.2011.5746332.

[21] S. Rai, J. Holleman, J. N. Pandey, F. Zhang, and B. Otis, "A 500µW Neural Tag with 2µVrms AFE and Frequency-Multiplying MICS/ISM FSK Transmitter", *2009 IEEE International Solid-State Circuits Conference – Digest of Technical Papers*, San Francisco, CA, 2009, pp. 212–213, 213a, doi: 10.1109/ISSCC.2009.4977383.

[22] J. H. Lee, W-J. Jung, J-W. Jung, J-E. Jang, and J. Park. "A Matched RF Charger for Wireless RF Power Harvesting System," *Microwave and Optical Technology Letters*, Vol. 57, 2015, doi: 10.1002/mop.29183.

Problems

1. What are the possible sources of energy harvesting?
2. Why is RF energy harvesting becoming popular?
3. What are the ranges of frequency of operations where RF harvesters can work satisfactorily?
4. What is the power density available in ambient from where RF energy can be harvested?
5. Why is the near-field communication technique less preferred for wireless power transfer?
6. What are the types of antennas used for RFEH systems?
7. What kind of polarization is used for antennas in RFEH systems?
8. What is a rectenna? What are its basic building blocks?
9. What are the bottlenecks of rectenna circuits?
10. What is power conversion efficiency (PCE)? What are the factors that influence the PCE?
11. Sketch a circuit diagram for a rectenna circuit used for 900 MHz.
12. What is the role of a diode in a rectenna circuit? How it can influence the frequency behavior of the circuit?
13. What is the role of an impedance matching network?
14. What are the different types of IMNs? Explain with schematics.
15. What is LNA? What are its specifications?
16. Why are voltage multiplier circuits necessary in rectenna modules?
17. What are the targeted applications of rectennas?

Future Scope

Printed antenna technology is not a new topic of discussion. Dates back, its theory was conceived and with the progress of fabrication technologies it has been implemented practically for various multifaceted applications. Currently it is finding wide applications in multiple fields of communication. As the "electronic eye" of a whole communication system, its design plays a key role. Future wireless communication is inevitable for ground-based applications, airborne usage, underwater applications, or use in the healthcare industry. Whether it is for consumer electronics or strategic applications, the future of printed antennas is very bright. An antenna engineer's life is becoming challenging day by day because of various multi-linked technological issues.

For future wireless communication, adaptive electronics is becoming a main area of research. In the field of antenna engineering, adaptivity or reconfigurability can be obtained either by altering the operating frequency or polarization, the radiation pattern, or a combination of more than one features. With the help of various switching circuits, these reconfigurable antennas can be realized for adaptive electronics.

Another important feature of modern-day electronics is flexible circuits. Very complicated circuits can be made compact and conformal in shape according to the requirements of practical applications where the end product may have a curved surface. Antennas fabricated on such flexible platforms are in high demand for space as well as for various crucial division applications.

In the upcoming era of the Internet of Things (IoT), sensors will be the main building blocks of whole networks, and wireless sensor modules (WSMs) will be the heart of such networks, where numerous sensors will be connected virtually by wireless means through defined protocols. Printed antennas will be a viable solution for this wireless communication.

Biomedical engineering will be another vast market for printed antenna engineering. Across the entire gamut of the EM spectrum, there are multiple applications for the welfare of humankind.

Last but not least is the integration capability of antenna modules with their associated driving electronics. This can be either in the form of a system on chip (SoC), system in package (SiP), or antenna in package (AiP). Antenna engineers also need to broaden their field to the packaging area for developing such systems in practice. This strategy will be very helpful for future 5G communication, automotive RADAR applications, 60 GHz-mmW applications, or to develop multi-input–multi-output (MIMO)–based wireless communication.

Appendices

Appendix 1

Table A.1 RF and Microwave Frequency Spectrum Along With Practical Applications

Band	Frequency Range	Application
Low frequency (LF)	30 to 300 kHz	Navigation, RFID, time standard, submarines
Medium frequency (MF)	300 kHz to 3 MHz	Marine/aircraft navigation, AM broadcast
High frequency (HF)	3 to 30 MHz	Mobile radio, short-wave broadcasting
Very high frequency (VHF)	30 to 300 MHz	Land mobile, FM/TV broadcast, amateur radio, aircraft communications
Ultra-high frequency (UHF)	300 MHz to 3 GHz	RFID, mobile phone, wireless LAN, PAN
L	1 to 2 GHz	Mobile satellite service (MSS), remote sensing, airborne warning and control (AWACS)
S	2 to 4 GHz	WiFi, Bluetooth, ZigBee biomedical
C	4 to 8 GHz	Satellite communication, WLAN, WiMax, WiFi and ISM band application
X	8 to 12 GHz	Radar, FSS family, police radar
Ku	12 to 18 GHz	VSAT, broadcast satellite Doppler navigation
K	18 to 26 GHz	Radar, satellite communication
Ka	26 to 40 GHz	Satellite communication
Q	30 to 50 GHz	Terrestrial microwave communication, radio astronomy
U	40 to 60 GHz	5G cellular, satellite communication
V	50 to 75 GHz	New WLAN, 802.11ad/WiGig
E	60 to 90 GHz	5G cellular
W	75 to 110 GHz	Automotive radar
F	90 to 140 GHz	Radar, radio astronomy
D	110 to 170 GHz	Millimeter wave, radar

Appendix 2

Table A.2 Comparison of Different Planar Antennas

Type	Radiation Pattern	Directivity	Bandwidth	Polarization
Microstrip	Broadside	Medium	Narrow	Linear/circular
Dipole	Broadside	Low	Medium	Linear
Monopole	Broadside	Low	Medium	Linear
PIFA	Broadside	Medium	Medium	Circular
Slot	Broadside	Medium/low	Medium	Linear
Bow tie	Broadside	Medium	Wide	Linear (vertical)
Circular loop	Broadside	Medium	Narrow	Linear/circular
Spiral	Broadside	Medium	Wide	Linear/circular
Quasi-Yagi	End fire	High	Wide	Linear
Fractal	Broadside	High	Wide	Linear/circular
Leaky wave	Scannable	High	Medium	Linear

Appendix 3

Table A.3 Comparison of Various Computational Electromagnetic Solvers

Parameter	MOM	FEM	FDTD
In-line method	Frequency domain	Frequency domain	Time domain
Advantage	Fast simulation	Flexible type with suitable mesh generation	Electrically large structure can be solved easily
Background mathematics	Integral equation	Differential equation	Electrically large structure can be solved easily
Best suited for	Electrically small antennas, wire antenna, planar antenna etc.	Arbitrary shapes, single or a band of frequencies	Electrically large structure, broadband
Operating principle	Frequency-dependent Green's function	Minimizing energy function	A discrete solution of Maxwell's equation

Appendix 4

Table A.4 Popular Types of Planar Antennas for Antenna-in-Package (AiP) Configuration

Antenna Type	Merits	Demerits	Usage (s)
Patch	• Compact • Light weight • Low profile • Can be multiband • Can be conformal • Polarization diversity	• Narrow BW • Low gain • Low power landline capability • Warpage	• Mobile phone • Base station • Imaging • RADAR
Yagi-Uda	• Compact • Light weight • Low profile • Wideband • High directivity	• Low power handing capability • Sensitive to its location on PCB • Non-polarization diversity	• Mobile phone
Grid	• High gain • Low X-polarization • Compact • Light weight • Wideband • Low profile	• Pattern squint • Low power • Narrow gain • Bandwidth	• Radar

Index

A

Antenna-in-Package (AiP), 73, 150
antipodal Vivaldi antenna, 206, 214
Apollonian fractal, 24, 84

B

Balun, 101, 214
band-notched UWB antenna
dual band, 54
multi-band, 55
single band, 48
bio-telemetry, 74
bow-tie antenna, 209
breast imaging, 205
bulk micromachining, 146

C

Cantor set, 16
capacitively loaded loop (CLL), 54
channel capacity loss (CCL), 160
chip-on-board (COB), 109
CMOS technology, 74
Cockcroft-Walton, 236
Cole-Cole model, 112
complementary split ring resonator
(CSRR), 58, 165
computational electromagnetics, 73
constant width slot antenna
(CWSA), 60
co-planar waveguide (CPW), 2, 23

D

defected ground structure (DGS), 161
diethylene glycol butyl ether, 125

dipole antenna, 2, 3
diversity gain (DG), 159
doublerigged horn, 219
dual band MIMO antenna, 194
dual band-notched UWB antenna, 54
dual elliptically tapered antipodal slot
antenna (DETSA), 62

E

electrical equivalent circuit, 122
electrically small antenna (ESA), 1, 73
electromagnetic band gap (EBG),
163, 184
electronic eye, 231
empirical modeling, 16, 141
envelope co-relation co-efficient
(ECC), 159
epoxy spreading, 150

F

FCC (Federal Commission
Communication), 35, 83
FDTD, 41, 246
FEM simulation, 82
fidelity factor, 206
figure-of-merit (FoM), 92
fractal printed antenna
Cantor set, 16
Hilbert curve, 19
Koch curve, 16
Minkowski, 19
Pythagoras, 19
Sierpinski carpet, 21
Sierpinski gasket, 18
Friis transmission formula, 115

250   Index

G

gastrointestinal (G.I) tract, 99
global positioning system (GPS), 178
ground penetrating radar (GPR), 160, 199

H

half annular-shaped antenna, 26, 88
Hammerstad-Bekkadal model, 142
high resistive silicon, 78
Hilbert curve, 19, 188
Hillocks, 141
Huray model, 143

I

IEEE802.11a, 46
IFA: inverted F-antenna, 5, 179
impedance matching network (IMN), 238
implantable medical devices (IMD), 196
IoT (Internet-of-Things), 225, 243
ISM band, 100, 116
iterative structure, 24
ITU, 55, 78

K

Kirchhoff's surface integral representation (KSIR), 41
Koch fractal, 20

L

LCP (liquid crystal polymer), 79, 129
leaky-wave antenna, 12
linearly tapered slot antenna (LTSA), 60
link margin, 126
log-periodic dipole antenna, 5
light emitting diode (LED), 107, 230
link-budget analysis, 126
long-term evolution (LTE), 155
line-of-sight (LOS), 238

M

Male Torso, 113
medical implant communication systems (MICS), 99
MEMS, 90, 91
meta-material structure (MTM), 170

micromachining, 91
microstrip leaky wave Antenna, 9
microstrip patch antenna, 2
microwave imaging, 22, 199, 172
MIMO technology, 155
miniaturized planar monopole antenna, 26
Minkowski fractal, 19, 118
monolithic microwave integrated circuits (MMICs), 40
monopole antenna, 157, 205
multi band MIMO antenna, 178
multichip modules (MCMs), 150
multiple band-notched UWB antenna, 55

N

near-field coupling (NFC), 238, 160
near-field imaging, 219
neutralization line (NL), 166

O

omega-shaped slots, 25

P

PEEK, 101, 125
Penta-Gasket-Koch (PGK), 23
perfect electric conductor (PEC), 107
polarization, 126
position emission tomography (PET), 200
potassium hydroxide (KOH), 146
power conversion efficiency (PCE), 228
printed antenna
    bow-tie antenna, 209
    dipole antenna, 3
    fractal antenna, 7
    inverted F-antenna, 5
    log-periodic antenna, 11
    microstrip antenna, 79, 134
    microstrip leaky wave antenna, 9
    monopole antenna, 20
    quasi-Yagi antenna, 6
    slot antenna, 38, 59
    spiral antenna, 8
Pythagoras tree fractal, 19

Q

Q-factor, 149
quasi-Yagi antenna, 6

Index 251

R

radio frequency energy harvesting (RFEH) technology, 225, 237
radio-frequency identification (RFID), 225
reconfigurable antenna, 73, 92
reconfigurable MIMO antenna, 188
rectangular microstrip antenna array (RMAA), 82, 134
rectenna, 225
RF energy harvester, 226, 237
RF harvesting, 226, 230
RFIC, 145
RF-MEMS, 90
RLC networks, 85, 141

S

self-similar property, 7, 25, 214
sensitivity, 226, 230
Shannon-Hartley theorem, 156
side lobe level (SLL), 60, 157
Sierpinski carpet, 17, 21
Sierpinski gasket, 17, 18
single input–signal output (SISO), 156
SIR, xxi
slot antenna, 36, 38
slot-line, 100, 101
smiley fractal antenna, 21
space-filling property, 25
specific absorption rate (SAR), 115, 206
spiral antenna, 8, 9
split ring resonator structure (SRR), 165
stiction, 148
sub-mmW, 7, 74, 79
substrate, 79, 84, 88
superstrate, 122, 123
super-wideband (SWB), 20, 74
surface micromachining, 91, 148
synthetic aperture radar (SAR), 199, 200, 220
System-in-Package (SiP), 150
system-on-chip (SoC), 133, 146

T

tapered slot antenna (TSA), 59, 206
    antipodal Vivaldi antenna, 64
    Vivaldi antenna, 62

tetra methyl amino hydroxide (TMAH), 146
TM10 mode, 75, 79
total active refection co-efficient (TARC), 159
through-wall imaging, 203

U

ULTRALAM, 133, 134
ultra-wideband MIMO antenna, 183
ultra-wideband (UWB) antenna, 35
    band notched UWB antenna, 48
    dual band notched UWB antenna, 54
    multiple band notched UWB antenna, 55
    printed monopole, 40
    single band notched UWB antenna, 48
    slot antenna, 52
UMTS, 179, 183

V

Vivaldi antenna, 206
VSWR, 21, 22, 24, 40
vector network analyzer (VNA), 112, 125
via-hole, 75, 145

W

wafer bonding, 148
weather radar, 203
wideband MIMO antenna, 179
WiFi, 35, 38, 155
WiMAX, 179, 245
wire bonding, 150
wireless body area network (WBAN), 42
wireless capsule endoscopy (WCE), 99, 121, 122
wireless local area network (WLAN), 245
wireless power transfer (WPT), 225, 226, 228
wireless sensor modules (WSMs), 243
wireless sensor network (WSN), 237

X

X-band, 58, 74, 76
X-ray, 200, 202, 220